Constraint Satisfaction Problems

To all my children, Emna, Zied, Molka and Skander.

Constraint Satisfaction Problems

CSP Formalisms and Techniques

Khaled Ghédira

Series Editor
Bernard Dubuisson

ISTE WILEY

First published 2013 in Great Britain and the United States by ISTE Ltd and John Wiley & Sons, Inc.

ISTE Ltd
27-37 St George's Road
London SW19 4EU
UK

www.iste.co.uk

John Wiley & Sons, Inc.
111 River Street
Hoboken, NJ 07030
USA

www.wiley.com

Library of Congress Control Number: 2012952416

British Library Cataloguing-in-Publication Data
A CIP record for this book is available from the British Library
ISBN: 978-1-84821-460-6

MIX
Paper from responsible sources
FSC® C013604
www.fsc.org

Printed and bound in Great Britain by CPI Group (UK) Ltd., Croydon, Surrey CR0 4YY

Table of Contents

Preface

This book is essentially based on a course offered for several years as part of the Master of Research in Sciences and Techniques for decision-making computing at the Higher Institute of Management, Tunis, and also of the Master of Research in Software Engineering and Decision-Support Systems at the National School of Computer Sciences, Tunis. It is also the reflection of research carried out in the *SOIE* (*Stratégies d'Optimisation et Informatique intelligentE* [1]) laboratory.

The book is intended for engineers, beginners or expert researchers, as well as teachers. For engineers it will make this field more accessible; for researchers it will outline basic notions and provide them with an extensive bibliography; and for teachers it will act as a course support.

First of all this book introduces the CSP formalism and its foundations in order to deal with Constraint Satisfaction Problems. Then it presents the main CSP based techniques which consist in either solving such problems by BT-like algorithms or speeding up this resolution by consistency enforcing and/or heuristics using and/or learning.

---

1 Optimization strategies and intelligent computing.

This book also reviews the extensions of CSP in terms of formalisms and techniques. A particular focus is given to the most known ones, namely Constraint Satisfaction and Optimization Problems (CSOP), Max-CSP and Distributed CSP.

The book is organized so as to be as most didactic possible. Indeed all the concepts and techniques are illustrated with simple examples. For more clarity, most of the algorithms are drawn on the example of the four-queens problem and the graph-coloring problem.

Introduction

The CSPs are ubiquitous in both the academic and industrial world and even in our everyday lives. The concept of constraint is a very old one. It has been used in many fields to represent various problems: textbook examples such as the n-queens game and the graph-coloring or cryptarithmetic problems, design issues such as VLSI, and industrial applications such as scene analysis and interpretation, planning, scheduling and allocation of resources.

Our aim is to express a problem as a set of variables and as a set of constraints on these variables. One of the most targeted objectives is the satisfaction of all the constraints (CSP). However, we will, in fact, see that many other objectives may be targeted. Many works have been carried out in this context, the main obstacle being combinatorics.

The first discipline to be interested in problems under constraints was indisputably operational research. Several models were developed; unfortunately, these were called into question owing to their difficulties in formalizing and solving real-life situations. In fact, operational research provides rigorous algorithmic techniques that are often unsuited to real-life problems: the difficulty of taking into account the specific characteristics of each problem, etc. New approaches usually placed under the banner of artificial intelligence (AI) helped to complete and improve these techniques, which gave impetus to the study of problems under constraints.

Among these approaches, we will focus in this book on the one that was developed around a simple and generic formalism: the CSP formalism. A CSP consists of a set of variables and a set of constraints. A finite and discreet domain of values is associated with each variable. A constraint is defined by a relation on a subset of the set of variables. We will go into more detail on this definition in Chapter 1. We can, however, on the basis of this definition, already target several objectives: the existence of a solution or not, producing a solution, producing all solutions, does a given variable's value belong to a given solution or all solutions, etc.

On the basis of this formalism and given the NP-complete aspect of this type of problem, a wide range of techniques had been developed by the artificial intelligence community: backtrack and its variants, coherence reinforcement or filtering, backtrack-filtering hybridization, etc. These techniques are grouped under the terms of approaches, methods or CSP techniques.

The CSP solving methods are generally categorized into two families: the first family includes the complete or the so-called exact methods, which ensure completeness. These are dedicated to strict CSPs, namely complete satisfaction, i.e. satisfaction of all the constraints. The second family includes incomplete or approximate methods that sacrifice quality in favor of time complexity and that can then obtain a "good" solution in a "reasonable" time but cannot, in practice, ensure completeness and/or optimality. Instead, they are dedicated to maximum constraint satisfaction problems (max-CSPs) and constraint satisfaction and optimization problems (CSOP), which are extensions of CSP. Hence, we devote an entire chapter to the basic CSPs, a chapter to max-CSPs, and another to CSOPs, to focus on the most appropriate solving methods.

Learning mechanisms were also integrated into these techniques to intelligently solve CSPs by storing information deemed useful for pruning the search space. We then devote a chapter to describe these mechanisms and the underlying algorithms in more detail.

Despite the wealth of literature, the basic CSP approach unfortunately still presents some limitations in particular at the level of the following points:

– *the formalism*: this essentially focuses only on constraints and not on the expression of preferences;

– *the resolution*: neither the previous reasoning nor the previous solution is reused to take into account, and optimally deal with, the dynamic aspect; the case of the problem changes linked to unforeseen events or imprecise knowledge;

– *the objectives*: the objectives are primarily looking for one or several solutions, the user's preferences for a particular solution not being possible.

Many studies have proposed extensions and brought enhancements to the basic CSP approach. They are particularly concerned with taking preferences, dynamicity and distribution into account.

As a result, other approaches such as distributed constraint satisfaction problems (DisCSPs) offer a new way to more efficiently apprehend the complexity of CSPs or the distributed aspect inherent to some real-life CSP applications. An extension is then proposed to give rise to the DisCSP formalism. This extension is discussed in more detail in Chapter 8.

This book is organized into eight chapters. The first chapter discusses the basic foundations of CSP formalism through a variety of definitions and examples of application. The second chapter is devoted to coherence reinforcement techniques. The third chapter describes the different methods for the complete resolution of the CSPs. The fourth chapter focuses on different heuristics and search approaches relating to CSPs to speed up the resolution and to better organize the route of the search space. In the fifth chapter, we draw up other approaches enhanced by storage and learning mechanisms.

Subsequently, we describe some extensions of CSPs. Thus, the sixth chapter discusses maximum constraint satisfaction problems (max-CSPs), while the seventh chapter discusses constraint satisfaction and optimization problems (CSOPs). Finally, we conclude the book with a chapter discussing an extension, based on the multi-agent approach, known as DisCSP.

Chapter 1

Foundations of CSP

Various real-life problems, considered as problems within artificial intelligence (AI) and operational research (OR), are solved via the constraint satisfaction problem (CSP) formalism that provides a tool for modeling and handling knowledge, and is simple, general, expressive and efficient in the sense that it is able to represent multiple and various types of knowledge. This formalism, introduced in the 1970s, is used in many domains, including task scheduling and the management and allocation of resources [GHÉ 06].

In this chapter, we begin by refining the definition of these CSPs as well as the associated basic notions. Then, we present the application areas of CSPs illustrated with examples prior to exposing the variants and extensions of the CSP formalism.

1.1. Basic concepts

DEFINITION 1.1:– VALUE.– *A value is something that can be assigned to a variable. Generally, we reference by v_i the value of the variable X_i. The nature of these values is a typification of the variables of the problem. When the values* *are numeric, the variables are said to be numeric. There are subtypes in this class of variables, such as integer, rational or real variables. Boolean variables can only take two possible values: true or false. There are also symbolic variables (e.g. the color that a variable can take in a graph-coloring problem).*

DEFINITION 1.2:– DOMAIN OF A VARIABLE.– *The domain of a variable is the set of all the values that this variable can take. If the variable is denoted by X_i, then the most general notation of the domain associated with this variable is either D_i or D_{xi}.*

DEFINITION 1.3:– DEGREE OF A VARIABLE.– *The degree of a variable is the number of constraints in which it is involved. Consider the following example:*

$X_1 + X_3 + 3X_2 < 15;$

$7X_2 * 4X_5 = 84;$

$2X_1 + 6X_4 - X_2 \geq 9X_3;$

Then, Degree (X_1) = 2, Degree (X_2) = 3, etc.

DEFINITION 1.4:– LABEL.– A label is a pair consisting of a variable and a value from its domain.

DEFINITION 1.5:– CONSTRAINT.– A constraint on a set of variables is a restriction on the set of values that these variables can take simultaneously.

DEFINITION 1.6:– ARITY OF A CONSTRAINT.– *The arity of a constraint C is the number of variables involved in C. A constraint is called unary if it relates to a single variable, but if its arity is equal to two, then we speak of a binary constraint. And more generally, a constraint is called n-ary if its arity is equal to n.*

DEFINITION 1.7:– INSTANTIATION.– *An instantiation* I *is the simultaneous assignment of values to a set of variables. This instantiation may be in the form of a set of values where each value relates to a variable; for example, the following tuple of values* $(v_1, v_2, \ldots, v_n)$ *is a possible instantiation of the variables* $(X_1, X_2, \ldots, X_n)$.

DEFINITION 1.8:– TOTAL/PARTIAL/CONSISTENT/INCONSISTENT INSTANTIATION.–

– A total instantiation is an instantiation of all the variables of the problem.

– A partial instantiation is an instantiation of a subset of variables.

– A total or partial instantiation is called consistent if and only if it satisfies all the constraints concerned by the variables that it involves.

– A total or partial instantiation is called inconsistent if the instantiation of some of its variables violates at least one constraint.

DEFINITION 1.9:– A NOGOOD.– *A Nogood is a partial instantiation that cannot be extended to a total consistent instantiation.*

1.2. CSP framework

The CSP formalism was introduced by Montanari in 1974 [MON 74]. It is informally defined by searching for the assignment of values to a set of variables submitted to certain conditions or constraints that restrict the choice of values.

1.2.1. *Formalism*

DEFINITION 1.10:– *CSP.– A CSP is a quadruplet* (X, D, C, R) *defined by*:

– a set of n variables: $X = \{X_1, X_2, \ldots, X_n\}$;

– a set of n finite and discreet domains: $D = \{D_1, D_2, \ldots, D_n\}$ where D_i is the set of possible values for the variable X_i such that $D = \{v_{i1}, v_{i2}, \ldots, v_{il}\}$ and $|D_i| = l$;

– a set of m constraints: $C = \{C_1, C_2, \ldots, C_m\}$ where any constraint C_i involves a subset of k variables $\{X_{i1}, X_{i2}, \ldots, X_{ik}\}$ such as $k \geq 1$;

– a set of m relations associated with the constraints where each one of the relations R_i defines all combinations of values permitted by C_i. R_i is defined by a subset of the Cartesian product of the k domains $D_{i1} \times D_{i2} \times \ldots \times D_{ik}$. This set is denoted by R such as $R = \{R_1, R_2, \ldots, R_m\}$.

A relation can be expressed in two ways:

– *In extension*: using a set of permitted tuples

Example: $R_1 = R_{X1\ X2} = \{(2, 1), (4, 2), (6, 3)\}$

– *In intention*: using characteristic functions or predicates

Example: $X_1 = 2X_2$

Solving a CSP amounts to finding a solution, i.e. assigning to each variable a value from its domain so as to satisfy all the constraints (see definition 1.11). Other issues may also arise: finding all the solutions, proving the existence of a solution, etc.

DEFINITION 1.11:– SOLUTION OF A CSP.– *A solution* S *of a CSP* (X, D, C, R) *is a consistent total instantiation*.

DEFINITION 1.12:– CONSISTENT CSP.– *A CSP is called consistent if and only if it possesses at least one solution.*

DEFINITION 1.13:– THE DIFFERENT REPRESENTATIONS OF A CSP.– *There are three possible representations for a CSP*:

– *Graphical representation*: This is the most commonly used. According to the level of detail desired, we can distinguish two different representations:

- either a representation via a graph local to the constraint (Figure 1.1),

- or a global representation via a graph of all CSP constraints, i.e. associating with any CSP (X, D, C, R) a graph of constraints $G = (X, C)$ whose nodes represent the variables and the edges represent the constraints.

– *Representation in extension*: The set of pairs authorized for the binary constraints or more generally the n-uplets authorized for the n-ary constraints.

– *Representation in intention*: The constraints are in the form of equations or predicates.

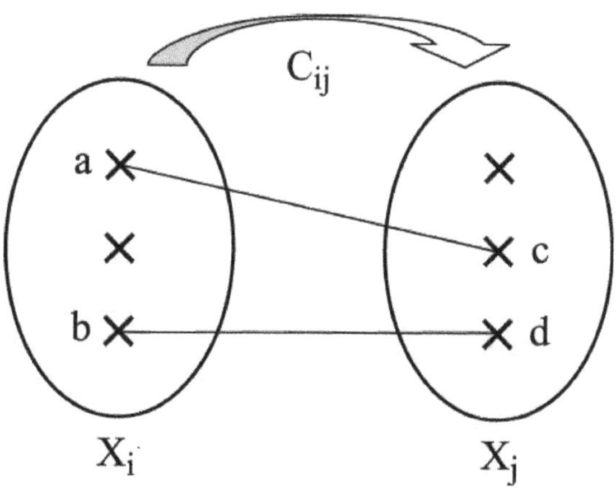

Figure 1.1. *Local graph of the constraint C_{ij}, linking X_i age and X_j*

1.2.2. *Areas of application*

Several problems of AI and/or combinatorial optimization and several industrial applications have been discussed in terms of CSP, such as scheduling [MIN 92], air traffic control [BAR 02], civil engineering [KAZ 08], mechanical engineering [YVA 08], cognition [VER 04], Web applications [BEN 04, BEN 06, MAR 06, LIN 05, RED 09, BOU 10], network security [BEL 06], protection of personal data or privacy [FAL 08], or even in the context of awareness [MOO 10].

Textbook cases or academic problems were also addressed by the CSP formalism. The most famous of these are the n-queens problem, the map-coloring problem, cryptarithmetic, etc.

In the following section, we describe examples of problems under constraints that have been formalized as CSPs.

EXAMPLE 1.1.– *The n-queens problem*

This problem consists of placing n-queens on an $n \times n$ chessboard so that no two queens attack each other. Thus, a solution requires that no two queens share the same row, column or diagonal. This example will serve as a basis to illustrate certain algorithms in the following chapters.

Because of the intrinsic symmetry of this problem and to reduce the combinatorics, we assume that each queen can only move along a single row exclusively reserved for it. Row i is associated with queen r_i. Because each queen r_i can only move on one of the squares from its row l_i, the constraint of the presence of a single queen per row is implicitly verified. The problem will be modeled as follows:

– $X = \{X_1, X_2, \ldots, X_n\}$, where X_i represents the number of the column where the queen associated with line i is placed.

– $D = \{D_1, D_2, \ldots, D_n\}$, where $D_i = \{1, 2, \ldots, n\}$.

– C is the following set of constraints:

- $\boldsymbol{C_1}$: $\forall\ i, j$: $X_i \neq X_j$, expressing that two queens must not be on the same column.

- $\boldsymbol{C_2}$: $\forall\ i, j$: $X_i + j \neq X_j + i$, expressing that two queens must not be in the same diagonal.

Figure 1.2 shows one solution to the four-queens problem.

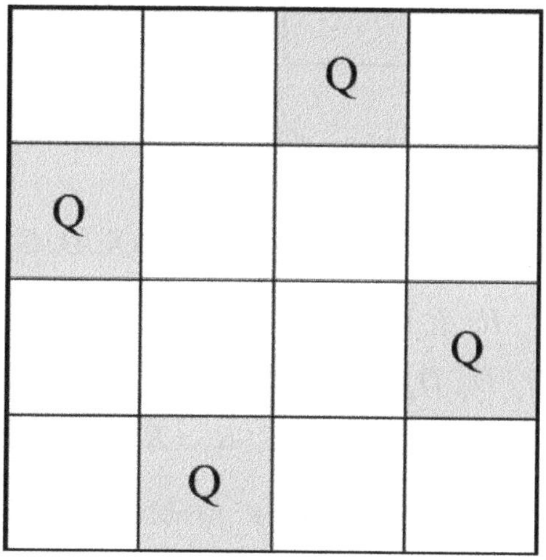

Figure 1.2. *A solution to the four-queens problem*

EXAMPLE 1.2.– *The map-coloring problem*

Another academic problem that can be formalized as a CSP is the map-coloring problem. This problem is a particular case of the graph-coloring problem. It is a matter of coloring a map with k colors in such a way that the two neighboring areas, having a common border, are not of the same color.

Figure 1.3 shows the map-coloring example with four regions to be colored with three colors (red, green and blue) as well as the associated graph of constraints.

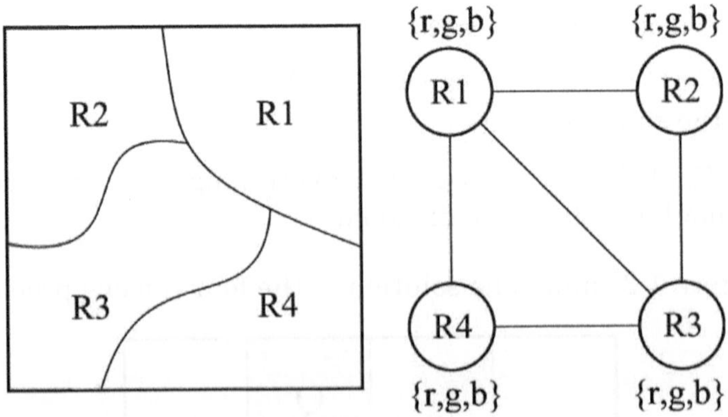

Figure 1.3. *Example of a chart-coloring problem*

The corresponding CSP is the triplet (X, D, C), such as:

– $X = \{R_1, R_2, R_3, R_4\}$.

– $D = \{D_1, D_2, D_3, D_4\}$, where $D_i = \{r, g, b\}$.

– $C = \{R_1 \neq R_2; R_1 \neq R_3; R_1 \neq R_4; R_2 \neq R_3; R_3 \neq R_4\}$.

A possible solution to this problem is illustrated in Figure 1.4.

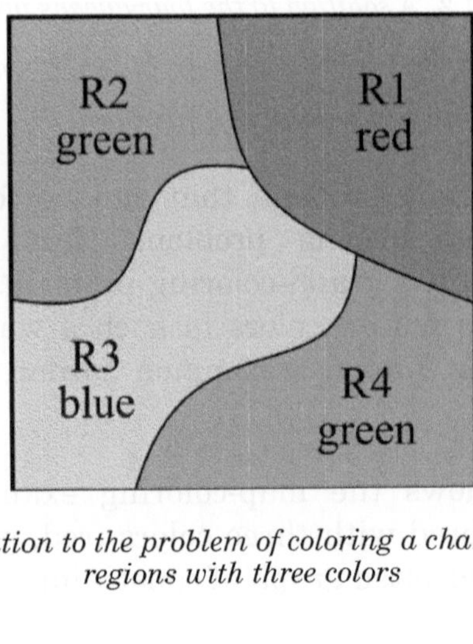

Figure 1.4. *A solution to the problem of coloring a chart consisting of four regions with three colors*

EXAMPLE 1.3.– *Cryptarithmetic example*

The most famous example is the following:

	S	E	N	D
+	M	O	R	E
-------------	-------	-------	-------	-------
= M	O	N	E	Y

The problem consists of finding a numerical value for each of the letters in such a way that the sum of "SEND" and "MORE" ends in the desired result, i.e. "MONEY".

– The variables of the associated CSP correspond to the different letters constituting the three nouns. Thus, the set X is as follows:

$$X = \{D, E, M, N, O, R, S, Y\}.$$

– The domains relating to these variables are as follows:

$$D_D = D_E = D_N = D_O = D_R = D_Y = \{0, 1, \ldots, 9\},$$

$$D_S = D_M = \{1, 2, \ldots, 9\}.$$

– The constraint C is defined as follows:

$$1{,}000 * S + 100 * E + 10 * N + D +$$

$$1{,}000 * M + 100 * O + 10 * R + E =$$

$$10{,}000 * M + 1{,}000 * O + 100 * N + 10 * E + Y.$$

We can obtain another model by adding four additional variables that represent the carry digits from adding the digits two-by-two. These variables are R_1, R_2, R_3 and R_4. Each one of them has the set {0, 1} as a domain. The model obtained is as follows:

– X = {D, E, M, N, O, R, S, Y, R_1, R_2, R_3, R_4};

– The domains relating to these variables are:

$D_D = D_E = D_N = D_O = D_R = D_Y = \{0, 1, \ldots, 9\}$,

$D_S = D_M = \{1, 2, \ldots, 9\}$,

$D_{R1} = D_{R2} = D_{R3} = D_{R4} = \{0, 1\}$.

– The constraints are:

C_1: $D + E = Y + 10 * R_1$

C_2: $N + R + R_1 = E + 10 * R_2$

C_3: $E + O + R_2 = N + 10 * R_3$

C_4: $S + M + R_3 = O + 10 * R_4$

C_5: $R_4 = M$

EXAMPLE 1.4.– *Unshareable resource allocation problem*

An unshareable resource allocation problem is defined by a set of resources and a set of tasks to be carried out. Each task t_i must, to be carried out, be assigned to one and only one resource r_j among a set of resources, and it consumes all its capacity [GHÉ 93].

Let the following assignment problem be:

– t_1 has the option to be placed on one of the resources r_3 or r_4.

– t_2 has the option to be placed on one of the resources r_1 or r_3.

– t_3 can be placed on r_2.

– t_4 has the option of being placed on one of the resources r_1, r_2 or r_4.

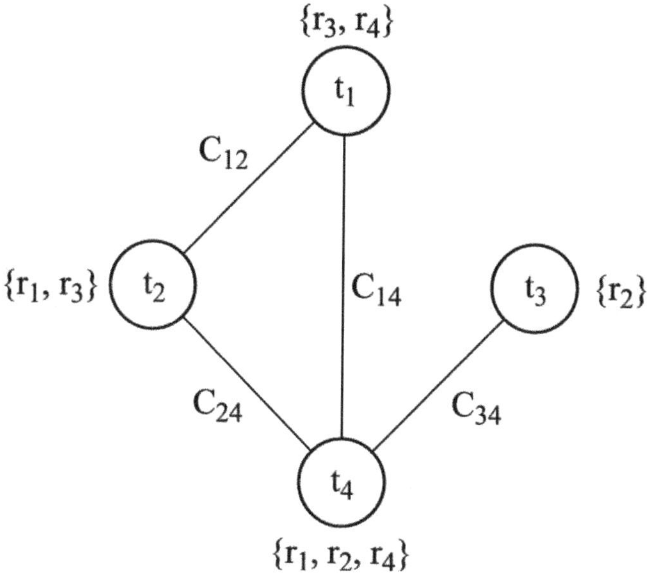

Figure 1.5. *The graph of constraints associated with the CSP of example 1.4*

The graph of constraints in Figure 1.5 represents the following CSP:

– The set of variables X is equal to $\{t_1, t_2, t_3, t_4\}$.

– The set of domains D is equal to $\{D_1, D_2, D_3, D_4\}$ with:

- $D_1 = \{r_3, r_4\}$
- $D_2 = \{r_1, r_3\}$
- $D_3 = \{r_2\}$
- $D_4 = \{r_1, r_2, r_4\}$

– The constraints on the variables are defined as follows:

- C_1 (associated with r_1), defined by $\{t_2, t_4\}$, is denoted by C_{24}.

- C_2 (associated with r_2), defined by $\{t_3, t_4\}$, is denoted by C_{34}.

- C_3 (associated with r_3), defined by $\{t_1, t_2\}$, is denoted by C_{12}.

- C_4 (associated with r_4), defined by $\{t_1, t_4\}$, is denoted by C_{14}.

– The relations, associated with the previously defined constraints, are represented in extension by the following authorized couples:

- $R_1 = \{(r_1, r_2), (r_1, r_4), (r_3, r_1), (r_3, r_2), (r_3, r_4)\}$, denoted by R_{24}.

- $R_2 = \{(r_2, r_1), (r_2, r_4)\}$, denoted by R_{34}.

- $R_3 = \{(r_3, r_1), (r_4, r_1)\}$, denoted by R_{12}.

- $R_4 = \{(r_3, r_1), (r_3, r_2), (r_3, r_4), (r_4, r_1), (r_4, r_2), (r_4, r_3)\}$, denoted by R_{14}.

EXAMPLE 1.5.– *Traveling salesman problem*

The example we deal with in this section is an instance of the well-known traveling salesman problem (TSP). A salesman (businessman here) has to visit several cities to meet customers in a given time interval. To do this, he already knows where he wants to leave from and where he wants to be at the end (cities of departure and arrival are not necessarily the same). As he has a lot of experience and knows many customers, he can predict the minimum amount of time he will spend in each city. In addition, he wants to go to some cities before others as he must see some customers before others. As he is in a hurry, he has decided to make all journeys by plane [LAW 85].

Let us assume that his customers are always available to him, regardless of time. Furthermore, let us assume that this salesman only wishes to take direct flights, i.e. he does not wish to change plane to reach a destination and he cannot visit the same city more than one time during one trip.

It is easy to see that the combinatorics of this problem increases exponentially with the number of cities.

Modeling this problem in the form of a CSP is as follows:

– All flights must satisfy the following constraints:

- $\boldsymbol{C_1}$**:** The trip must take place within a given time interval $[t_1, t_2]$.

- $\boldsymbol{C_2}$**:** The first flight must leave the departure city of the trip, $Start_{city}$.

- $\boldsymbol{C_3}$**:** The last flight must land in the arrival city of the trip, End_{city}.

- $\boldsymbol{C_4}$**:** It is necessary to go to each city and stay there for a minimum duration. Let m be the number of cities to be visited, V the set of cities in question and *Dur* the set of minimal time duration spent in each one of these cities.

- $\boldsymbol{C_5}$**:** The precedences between the cities to be visited must be respected. That is to say the function *Pred* (v_1, v_2) that returns true if and only if the city v_1 is visited before v_2.

- $\boldsymbol{C_6}$**:** The salesman takes only direct flights.

– Let n be the number of flights to be taken.

We will model our problem with flight variables f_i, $I = 1, \ldots, n$, where f_i represents the i-th flight of the trip. The domain of the values is the same for all the f_i; that is to say all of the cities this salesman needs to visit. In doing so, we directly comply with constraint C_6. Now we need to formalize the other constraints using the variables f_i.

We define the constraint C_1 as:

$DepartureDate\ (f_1) \geq t_1$,

$ArrivalDate\ (f_n) \leq t_2$,

which requires that every flight takes place within a given time interval:

$$\forall i \in [1...n] \begin{cases} \textbf{DepartureDate}(f_i) \geq t_1 \\ \textbf{ArrivalDate}(f_i) \leq t_2 \end{cases}$$

Constraints C_2 and C_3 are given by the following equations:

$$\text{DapartureCity}(f_1) = \text{city}_{\text{start}}$$

$$\text{ArrivalCity}\ (f_n) = \text{city}_{\text{end}}$$

The constraint C_4 results in the following four equations, where *Interval* (a, b) is the time elapsed between the arrival of the flight a and the departure of the flight b. Note that the fourth equation ensures that all the cities have to be visited:

$$\forall i \in [1...n]\ \textbf{DepartureCity}(f_i) \in V$$

$$\forall i \in [1...n],\quad \textbf{ArrivalCity}(f_i) \in V$$

$$\forall i \in [1...n-1], \begin{cases} \textbf{ArrivalCity}(f_i) = \textbf{DepartureCity}(f_{i+1}) \\ Interval(f_i, f_{i+1}) \geq Dur(\textbf{ArrivalCity}(f_i)) \end{cases}$$

$$\forall i \neq k \in [1...n],\quad \textbf{ArrivalCity}(f_i) \neq \quad \textbf{ArrivalCity}(f_k)$$

Formulating the constraint C_5 using only f_i variables is quite tedious. We will introduce new variables t_j, one for each city j to be visited, to simplify our task. We will call them the "city" variables (as opposed to the "flight" variables). Each variable t_j indicates the order in which the city j is visited: if $t_j = p$, this will mean that j is the p-th city of the trip. The domain of values for these variables is therefore the progression 1, ..., m (m being the number of cities to be visited). C_5 is now very simply described by the equation:

$$\forall j \neq k \in V, \quad t_j < t_k \text{ } \mathit{if} \ \mathit{predecessor}(j,k)$$

It is, however, necessary to connect the two types of variables, which is achieved via the equation:

$$\forall i \in [1...n], \forall j \in V \begin{cases} i = t_j \ \mathit{if} \ \mathit{DepartureCity}(f_i) = j \\ i \neq t_j \ \mathit{if} \ \mathit{DepartureCity}(f_i) \neq j \end{cases}$$

EXAMPLE 1.6.– *Euler's knight problem*

This problem is much discussed in many books. It consists of moving a knight on a $n \times n$ chessboard in such a way that it goes to each square once and once only and at the end returns to the departure square.

EXAMPLE 1.7.– *The Shurr problem*

This problem consists of depositing n marbles numbered from 1 to n in m boxes. The constraints to be satisfied are as follows: a single box must not contain two marbles x and y such as $x = 2^* y$, nor three marbles x, y, and z such as $x + y = z$. The complexity of this problem increases exponentially with the size n.

EXAMPLE 1.8.– *The fault diagnosis problem*

This problem involves making several initial observations, and then making assumptions called "scenarios" and finally choosing a scenario from these scenarios. For a scenario to be a possible solution, it must satisfy all the constraints that represent the functional model of the object we wish to analyze (machine, human body, etc.).

EXAMPLE 1.9.– *The car-sequencing problem*

This problem is considered to be one of the most significant problems in the automobile production industry. Indeed, after constructing the basic model of a car, several options (e.g. air conditioning, metallic paint and ABS brakes) are added. Different models of cars require different combinations of options. The cars are placed on a moving assembly line and pass by numerous workstations where these options are installed. The constraints relate to the total amount of time spent by cars in a single workstation, the installation time of each option and the capacity of each of the workstations.

Given a set of n cars, each one assuming a given set of options, the problem involves moving the cars onto the assembly line without exceeding any of the service station's capacities. Other constraints can be added to the real problem such as the start date, end date and the just in time.

EXAMPLE 1.10.– *The problem of frequency assignment for cellular systems*

It is the dramatic increase in the number of demands for mobile telephone services, which is the origin of the importance of this problem. There is a set of geographically divided hexagonal regions called "cells". Frequencies must be assigned to the cells depending on the number of calls and the geographical positions of the cells.

The objective is to find a frequency assignment that satisfies the constraints by using the minimum number of frequencies (constraint on the maximum number of frequencies available).

1.2.3. *Extensions*

Several extensions of the CSP formalism are proposed in the related literature. They aim to expand the scope of this formalism by introducing new concepts. Among these extensions, we include the following.

1.2.3.1. *Dynamic CSP*

The majority of problems, especially real-life problems, evolves over time due to the environment and the user: change in objective, presence of hazards (weather and faults), interactive system, etc.

To be able to solve problems of this type, we must find incremental algorithms that are capable, if the problem changes, of producing a new result (as quickly as possible) that remains to be of a "good" quality. It is therefore important that the new solution produced is not too remote, according to a measure that is essentially dependent on the problem from the preceding solution, the property of stability.

In terms of CSP, the changes can be of different types: adding or withdrawing variables, values, constraints and tuples of values in the relations associated with the constraints. But all these changes in fact amount to either adding or removing constraints. This is how the dynamic CSPs (DCSPs) were defined [DEC 88]:

A DCSP is defined by a sequence P_0, ..., P_k of CSPs. Each CSP P_{i+1} is the result of applying an elementary modification to the problem P_i. This modification may be a restriction (addition of constraint(s)) and/or a relaxation (removal of constraint(s)).

The removal of a constraint preserves the solutions, but calls into question all the deductions and particularly any filtering that may have occurred. Adding a constraint preserves the deductions and, therefore, the effects of filtering, but calls the solutions into question. Note that filtering is a mechanism that can be applied before seeking solutions and corresponds to the removal of parts of the problem to achieve an equivalent problem, but after verifying a certain degree of consistency.

1.2.3.2. *Domain hierarchical CSP*

The domain hierarchical CSP (DHCSP) is a CSP in which a partial order (a hierarchy of specialization/generalization) is defined for all the domains of variables in such a way that these hierarchies preserve compatibility, i.e. if two elements a and b having different domains are compatible then the specializations of these two elements are also compatible. A DHCSP is a natural formalization of constraints for the problem defined in an object-oriented environment where a partial order (inheritance hierarchy) is defined for the objects. This hierarchy can be used to improve the handling of constraints.

1.2.3.3. *Distributed CSP*

In this formalism, the CSP is distributed among several agents. Each agent handles a sub-CSP. The aim is then to solve all the sub-CSPs by satisfying the inter-agent constraints. More details will be provided in Chapter 8.

1.2.3.4. *Constraint satisfaction and optimization problem*

This formalism [HAO 99] incorporates an objective function. The aim is then not only to satisfy all the constraints but also to optimize this objective function. More details will be provided in Chapter 7. There is also a version, called the constrained optimization problem (COP), more flexible than the constraint satisfaction and optimization problem (CSOP), which consists of relaxing certain constraints and/or penalizing the violated constraints.

1.2.3.5. *Constraint satisfaction and multi-criteria optimization problem*

This formalism, proposed by [BEN 07], extends the CSOP to the case of an objective function with several criteria. The aim is not only to satisfy all the constraints but also to optimize several criteria at once. This optimization, in fact, amounts to finding the best possible compromise among these criteria. Several real-life problems were formalized and dealt with as a constraint satisfaction and multi-criteria optimization problem (CSMOP), including the problem of road traffic regulation within multimodal urban transport [BOU 08, BOU 09].

1.2.3.6. *Partial CSP*

One of the major limitations of the standard CSP formalism, requiring total satisfaction, i.e. of all the constraints, is the lack of flexibility while dealing with over-constrained problems that result, in the majority of cases, in the absence of a solution. [FRE 92] proposed a generic extension based on the notion of relaxing the original over-constrained problem to give an equivalent problem that is simpler and above all solvable.

Relaxation of a problem can be obtained by removing a variable, withdrawing a constraint and extending the domain of a variable or extending the domain of a relation. Hence, the notion of a partial constraint satisfaction problem (PCSP).

A PCSP is a triplet $<$ (P, (PS, $\leq$)), (M, (N, S)) $>$ such that:

– P is the initial CSP,

– (PS, $\leq$) is the space of partially ordered P_i problems containing P such that PS is a set of CSP P_i and "$\leq$" is a partial order on these problems. $P_1 \leq P_2$ is verified if the set of P_2 solutions is a subset of the set of P_1 solutions, i.e. Sols $(P_2) \subseteq$ Sols (P_1).

– M is a metric (distance function) on (PS, ≤), and (N, S) are the necessary and sufficient boundaries of the distance between the original problem P and a solvable problem P' in PS.

A solution to a PCSP is given by a P' problem from the PS problem and its solutions, such as the distance between P and P' is less than N. If the distance between P and P' is minimal then we have an optimal solution.

1.2.3.7. *Maximal CSP*

Maximal CSP (Max-CSP) is a variant of the PCSP, which aims to find a complete instantiation that maximizes the number of satisfied constraints or, equivalently, which minimizes the number of constraints that are violated. More details will be provided in Chapter 6.

1.2.3.8. *Valued CSP*

PCSP and Max-CSP were the first serious attempts to meet the challenge of handling over-constrained problems. A subsequent reading of this scenario suggested the distinction between two types of constraint, which can coexist in the same problem: the hard constraints that reflect obligations to be imperatively satisfied and the soft constraints that represent preferences, i.e. where possible they should not be violated. This reading has given rise to several specific extensions, which are generally based on the relaxation of constraints such as weighted CSPs, fuzzy constraint satisfaction problems (Fuzzy CSP) [DUB 93, ROS 76, RUT 94], possibilistic constraint satisfaction problems (Possibilistic CSP) [SCH 92] and probabilistic constraint satisfaction problems (Probabilistic CSP) [FAR 93]. Given the obvious similarity between these extensions, generic frameworks were proposed, particularly, the valued CSP (VCSP) [SCH 95] that includes most of these extensions. This framework defines valuations on the constraints using an ordered commutative monoid, i.e. an ordered set of valuations with an aggregation operator satisfying certain properties.

Formally, a VCSP is based on an algebraic structure called *valuation*, linked to a monoid (E, ⊛, ≻, ⊥, T) such that:

– E is a set of valuations taken in the interval [0, 1], totally ordered by ≻,

– ⊥, T represent the minimum and maximum of E, respectively,

– ⊛ is an associative, commutative and binary internal operation on E, which verfies:

- neutral element: **∀ a ∈ E, (a ⊛ ⊥) = a.**

- absorbing element: **∀ a ∈ E, (a ⊛ T) = T.**

- monotonicity: **∀ a, b, c ∈ E, (a ≽ b) ⇒ ((a ⊛ c) ≽ (b ⊛ c)).**

By selecting a particular valuation structure, as Table 1.1 shows, we may end up with one of the specific extensions previously mentioned. Given that these valuations are associated with the constraints, FCSP cannot be instantiated from a VCSP since it evaluates the instantiations instead of the constraints.

	E	⊛	≻	⊥	T
standard CSP	{true, false}	∧	>	true	false
possibilistic CSP	[0, 1]	*max*	>	*0*	*1*
weighted CSP	ℕ ∪ {+∞}	+	>	*0*	+∞
probabilistic CSP	[0, 1]	×	<	*1*	*0*
lexicographical CSP	$\mathbb{N}^{[0,1]}$ ∪ {T}	∪	$>_{lex}$	∅	T

Table 1.1. *The instantiations of the VCSP according to the chosen valuation structure*

Finally, it is important to mention another generic framework similar to VCSP called semiring-based CSP (SCSP) [BIS 95]. A detailed comparison between VCSP and SCSP is provided in [BIS 96] and [BIS 99].

Several other works were proposed to extend and/or complete the standard CSP formalism. We quote mixed CSP [FAR 96], adaptive CSP [BOR 96], random CSP [ACH 97, GEN 01, MAC 98], structural CSP [NAR 99], quantified CSP [BEN 00], stochastic CSP [WAL 02], open CSP [FAL 02], uncertain CSP [YOR 03], ordinal CSP [FRE 03], fuzzy CSP [LIN 05] and decentralized CSP [DUF 11].

1.3. Bibliography

[ACH 97] ACHLIOPTAS D., KIROUSIS L.M., KRANAKIS E., KRIZANC D., MOLLOY M.S.O., STAMATIOU Y.C., "Random constraint satisfaction: a more accurate picture", *Proceedings of International Conference on Principles and Practice of Constraint Programming*, LNCS, Springer, pp. 107–120, 1997.

[BAR 02] BARNIER N., Application de la programmation par contraintes à des problèmes de gestion du trafic aérien, PhD from INP, Toulouse, France, 2002.

[BEL 06] BELLA G., BISTARELLI, S., FOLEY, N., "Soft constraints for security", *Proceedings of 1st International Workshop on Views on Designing Complex Architectures (VODCA'04) in Electronic Notes in Theoretical Computer Science (ENTCS)*, vol. 142, pp. 11–29, 2006.

[BEN 04] BENBERNOU S., CANAUD E., PIMONT S., *Semantic WS Discovery Regarded as a Constraint Satisfaction Problem*, LNCS, vol. 3055, pp. 282–294, 2004.

[BEN 00] BENHAMOU F., GOUALAR F., "Universally quantified interval constraints", *Proceedings of International Conference on Principles and Practice of Constraint Programming*, LNCS, Springer, pp. 67–82, 2000.

[BEN 06] BEN HASSINE A., MATSUBARA S., ISHIDA T., "A constraint based approach to horizontal web service composition", *Proceedings of International Semantic Web Conference*, LNCS, pp. 130–143, 2006.

[BEN 07] BEN JAÂFAR I., GHÉDIRA K., BOUDALI I., "Coordination based multiple criteria decision making", *Journal of Decision Systems*, vol. 16, no. 1, pp. 37–55, 2007.

[BIS 95] BISTARELLI S., MONTANARI U., ROSSI F., "Constraint solving over semirings", *Proceedings of International Joint Conference on Artificial Intelligence*, San Francisco, USA, pp. 624–630, 1995.

[BIS 96] BISTARELLI S., FARGIER H., MONTANARI U., ROSSI F., SCHIEX T., VERFAILLIE G., "Semiring-based CSP and valued CSP: basic properties and comparison", in JAMPEL M., FREUDER K., MAHER M. (eds), *Over-Constrained Systems*, Springer-Verlag, LNCS 1106, pp. 111–150, 1996.

[BIS 99] BISTARELLI S., MONTANARI U., ROSSI, SCHIEX T., VERFAILLIE G., FARGIER H., "Semiring based CSP and valued CSP: frameworks, properties, and comparison", *Constraints Journal*, vol. 4, no.3, pp. 199–240, 1999.

[BOR 96] BORRETT J.E., TSANG E.P.K., WALSH N.R., "Adaptive constraint satisfaction", *Proceedings of the UK Planning and Scheduling workshop*, Liverpool, UK, 1996.

[BOU 08] BOUDALI I., BEN JAÂFAR I., GHÉDIRA K., "Distributed decision evaluation model in public transportation systems", *International Journal of Engineering Applications of Artificial Intelligence*, vol. 21, no. 3, pp. 419–429, 2008.

[BOU 09] BOUDALI I., GHÉDIRA K., "A distributed multicriteria approach for traffic regulation in public transport systems", *International Journal of Applied Artificial Intelligence*, Taylor & Francis, vol. 23, no. 7, pp. 599–632, 2009.

[BOU 10] BOUSTIL A., SABOURET N., "Towards a semantic approach for WS composition handling users constraints", *Proceeding of International conference on Information Integration and Web-based Applications & Services*, Paris, France, 2010.

[DEC 88] DECHTER R., DECHTER A., "Belief maintenance in dynamic constraint networks", *Proceedings of National Conference on Artificial Intelligence*, St. Paul, Minnesota, USA, AAAI Press, pp. 37–42, 1988.

[DUB 93] DUBOIS D., FARGIER H., PRADE H., "Using fuzzy constraints in job-shop scheduling", *Proceedings of the Workshop on Knowledge-based Production Planning, Scheduling and Control, in conjunction with International Joint Conference on Artificial Intelligence*, Chambéry, France, 1993.

[DUF 11] DUFFY K.R., BORDENAVE C., LEITH D.J., "Decentralized constraint satisfaction", *The Computing Research Repository (CoRR)*, 2011.

[FAL 08] FALTINGS B., LÉAUTÉ T, PETCU P., "Privacy guarantees through distributed constraint satisfaction", *International Conference on Web Intelligence and Intelligent Agent Technology 2008 IEEE/WIC/ACM*, 9–12 December, Sydney, NSW, Australia, 2008.

[FAL 02] FALTINGS B., MACHO-GONZALEZ S., "Open constraint satisfaction", *Proceedings of the International Conference on Principles and Practice of Constraint Programming (CP'02)*, LNCS, Springer-Verlag, vol. 2470, pp. 356–370, 2002.

[FAR 93] FARGIER H., LANG J., "Uncertainty in constraint satisfaction problems: a probabilistic approach", *Proceedings of European Conference on Symbolic and Quantitative Approaches to Reasoning and Uncertainty*, LNCS, Springer-Verlag, vol. 747, pp. 97–104, 1993.

[FAR 96] FARGIER H., LANG J., SCHIEX T., "Mixed constraint satisfaction: a framework for decision problems under incomplete knowledge", *Proceedings of the 12th National Conference on Artificial Intelligence, Portland, Oregon,* AAAI Press, pp. 175–180, 1996.

[FRE 92] FREUDER E.C., WALLACE R.J., "Partial constraint satisfaction", *Artificial Intelligence*, vol. 58, nos. 1–3, pp. 21–70, 1992.

[FRE 03] FREUDER E.C., WALLACE R.J., HEFFERNAN R., "Ordinal constraint satisfaction", *Proceedings of the 5th International Workshop on Soft Constraints, in conjunction with International Conference on Principles and Practice of Constraint Programming*, LNCS, Springer, 2003.

[GEN 01] GENT I., MACINTYRE E., PROSSER P., SMITH B., WALSH T., "Random constraint satisfaction: flaws and structure", *Constraints*, vol. 6, no. 4, pp. 345–372, 2001.

[GHÉ 93] GHÉDIRA K., MASC une approche Multi-Agents des Problèmes de Satisfaction de Contraintes, PhD Thesis ENSAE, Toulouse, France, 1993.

[GHÉ 06] GHÉDIRA K., *Logistique de la production: approches de modélisation et de résolution*, Editions TECHNIP, 2006.

[HAO 99] HAO J., GALINIER P., HABIB M., "Métaheuristiques pour l'optimisation combinatoire et l'affectation sous contraintes", *Revue d'Intelligence Artificielle*, vol. 13, no. 2, pp. 283–324, 1999.

[KAZ 08] KAZUNORI M., YUKIO F., SEIICHI N., "Urban traffic signal control based on distributed constraint satisfaction", *Proceedings of the Annual Hawaii International Conference on System Sciences*, Waikoloa, Hawaii, 2008.

[LAW 85] LAWLER E.L., LENSTRA J.K., RINNOOY KAN A.H.G., SHMOYS D.B., *The Traveling Salesman Problem*, John Wiley, Chichester, 1985.

[LIN 05] LIN M., XIE J., GUO H., WANG H., "Solving QoS-driven WS dynamic composition as Fuzzy constraint satisfaction", *Proceedings of IEEE International Conference on E-Technology, E-commerce and EService*, Hong Kong, China, 2005.

[MAC 98] MACINTYRE E., PROSSER P., SMITH B., WALSH T., "Random constraint satisfaction: theory meets practice", *Proceedings of Principles and Practices of Constraint Programming (CP'98)*, Springer, pp. 325–339, 1998.

[MAR 06] MARUYAMA D., PAIK I., SHINOZAWA M., "A flexible and dynamic CSP solver for WS composition in the semantic web environment", *Proceedings of the 6th IEEE International Conference on Computer and Information Technology*, Seoul, Korea, p. 43, 2006.

[MIN 92] MINTON S., JOHNSON M.D., PHILIPS A.B., LAIRD P., "Minimizing conflicts: a heuristic method for constraint-satisfaction and scheduling problems", *Artificial Intelligence*, vol. 58, pp. 161–205, 1992. [Reprinted in *Constraint-Based Reasoning*, FREUDER E.C., MACKWORTH A.K. (eds), MIT Press, 1994.]

[MON 74] MONTANARI U., "Networks of constraints: fundamental properties and applications to picture processing", *Information Sciences*, vol. 7, pp. 95–132, 1974.

[MOO 10] MOORE P., JACKSON M., HU B., "Constraint satisfaction in intelligent context-aware systems", *Proceedings of International Conference on Complex, Intelligent and Software Intensive Systems*, Krakow, Poland, pp. 75–80, 2010.

[NAR 99] NAREYEK A., "Structural constraint satisfaction", Papers from the *1999 AAAI Workshop on Configuration*, Technical Report, WS-99-05, AAAI Press, Menlo Park, CA, pp.76–82, 1999.

[RED 09] REDDY S., GAL Y., SHIEBER S., *Recognition of Users' Activities Using Constraint Satisfaction*, Springer Berlin/Heidelberg, vol. 5535, pp. 415–421, 2009.

[ROS 76] ROSENFELD A., HUMMEL R., ZUCKER S. "Scene labeling by relaxation operations", *IEEE Transactions on* Systems, *Man, and Cybernetics*, vol. 6, no. 6, pp. 173–184, 1976.

[RUT 94] RUTTKAY Z., "Fuzzy constraint satisfaction", *Proceedings of the 3rd IEEE International Conference on Fuzzy Systems*, IEEE Press, pp.1263–1268, 1994.

[SCH 92] SCHIEX T., "Possibilistic constraint satisfaction problems or 'How to handle soft constraints?", *Proceedings of the Eighth Annual Conference on Uncertainty in Artificial Intelligence*, Stanford University, Stanford, USA, pp. 268–275, 1992.

[SCH 95] SCHIEX T., FARGIER H., VERFAILLIE G., "Valued constraint satisfaction problems: hard and easy problems", *Proceedings of the 14th International Joint Conference on Artificial Intelligence*, Montréal, Canada, pp. 631–637, 1995.

[VER 04] VERA A., HOWES A., MCCURDY M., LEWIS R.L., "A constraint satisfaction approach to predicting skilled interactive cognition", *Proceedings of the Conference on Human factors in Computing Systems*, New York, pp. 121–128, 2004.

[WAL 02] WALSH T., "Stochastic constraint programming", *Proceedings of the European Conference on Artificial Intelligence*, IOS Press, Amsterdam, The Netherlands, 2002.

[YOR 03] YORKE-SMITH N., GERVET C., "Certainty closure: a framework for reliable constraint reasoning with uncertainty", *Proceedings of the CP'03*, LNCS, vol. 2833, pp. 769–783, 2003.

[YVA 08] YVARS P-A., "Using constraint satisfaction for designing mechanical systems", *International Journal on Interactive Design and Manufacturing (IJIDeM'08)*, vol. 2, no. 3, pp. 161–167, 2008.

Chapter 2

Consistency Reinforcement Techniques

Consistency reinforcement techniques, also called filtering techniques, aim to simplify the resolution of a CSP before and/or during the solution search by providing as much information as possible. In concrete terms, they reduce the domains of variables and/or relations associated with the constraints by respectively removing values and/or n-uplets of values that cannot appear in any solution. They also allow for the early detection of failures. However, we must look for the best possible compromise between the time required for these reductions and the quantity of information removed. In this chapter, we discuss the different consistency reinforcement properties as well as the algorithms that allow them to be established.

2.1. Basic notions

2.1.1. *Equivalence*

Two CSPs, P and P', are called equivalent if they have the same set of solutions. A CSP P' problem is called simpler than a CSP P if P' is obtained from P by reducing domains and/or relations associated with P. If after this filtering, the domains of the variables and/or the relations associated with the constraints of P are empty then P has no solution.

2.1.2. *K-consistency*

DEFINITION 2.1:– K-CONSISTENCY.– *A CSP (X, D, C, R) is k-consistent if and only if, for any n-uplet of k variables (X_1, ..., X_k) of X, any consistent k − 1 instantiation may be extended to a consistent instantiation with the k-th variable.*

The *k-consistency* property of a CSP does not necessarily imply its consistency, i.e., this CSP does not necessarily have a solution even it is *k-consistent*. Besides if a CSP containing *k* variables is strongly *k-consistent* (see next definition) then it is consistent. [BES 01] shows that *k-consistent* CSP is not necessarily (*k* − 1) consistent and defines a stronger property: *strong k-consistency*.

DEFINITION 2.2:– STRONG K-CONSISTENCY.– *A CSP P (X, D, C, R) is said to be strongly k-consistent if and only if, $\forall i$, $1 \leq i \leq k$, P is i-consistent.*

A general algorithm that can be used to establish a strong *k*-consistency was proposed in [TSA 93]. Applied to a CSP of *n* variables whose maximum domain size is equal to *d*, the spatial and temporal complexity of this algorithm is in $O(n^k d^k)$. As a result, the most frequently used filtering algorithms are those that establish a low-level consistency.

DEFINITION 2.3:– NODE CONSISTENCY.– *A node-consistent CSP (X, D, C, R) is a 1-consistent CSP. This consistency is only verified if for any X_i variable of X, and for any v_i value of D_i, the partial assignment (X_i, v_i) satisfies all the unary constraints of C involving this variable.*

Node consistency is easy to verify since we can remove all the values that do not satisfy any of these unary constraints from the domain of each variable. This consistency is often taken for granted. For a problem P possessing n variables and whose maximum domain size is d, the node consistency is obtained in $O(nd)$.

DEFINITION 2.4:– ARC CONSISTENCY.– A CSP (X, D, C, R) is called arc consistent *if and only if, for any couple of variables* (X_i, X_j) *of* X*, each couple represents an arc in the associated constraint graph, and for any value* v_i *from the domain* D_i *that satisfies the unary constraints involving* X_i*, there is a value* v_j *in the domain* D_j *compatible with* v_i.

Initially presented by A.K. Mackworth [MAC 77], the arc consistency is expressed on each couple of variables of a problem with binary constraints. It is equivalent to the 2-consistency.

Other forms of local consistency have also been defined. These include:

– Path consistency (PC) [ALL 94, MAC 77, TSA 93], which, in the case of binary CSPs, corresponds to strong 3-consistency. A CSP is called path consistent (i.e. 3-consistent) if and only if, for any triplet, any consistent instantiation with two variables can be extended to a consistent instantiation with the third. A CSP is called strongly path consistent if and only if it is both arc consistent and path consistent. According to [ALL 94], this consistency is of a great complexity, but it remains interesting for certain types of problem for which the strong 3-consistency ensures global consistency. Several algorithms were proposed to establish path consistency. These include: PC1 [MON 74], PC2 [MAC 77], PC3 [MOH 86], PC4 [HAN 88], PC8 [CHM 98], PC2001 [BES 05] and DC2 [LEC 07a].

– Directional arc consistency and path consistency [TSA 93], which limit the local consistency properties when filtering, are used prior to seeking a solution via the backtracking algorithm and using a static vertical order (see section 4.2.2).

– The (i, j) consistency was introduced by Freuder in 1985 and extensively detailed in [TSA 93]. It corresponds to a generalization of the k-consistency. However, it remains of a largely theoretical interest.

2.2. Arc consistency reinforcement algorithms

The majority of works proposed within the field of consistency reinforcement or filtering are interested in arc consistency. We quote the best known that are denoted by AC-i. The earliest algorithms – AC-1 [MAC 77, ROS 76, TSA 93], AC-2 [MAC 77] and AC-3 [MAC 77, TSA 93] – rely on a very simple function called *Revise*, which is applied to a couple of variables (X_i, X_j) connected by a constraint C_{ij} by removing the locally inconsistent values from the X_i domain. This couple of variables represents an arc in the graph of constraints often denoted by (i, j). The arc consistency is verified if and only if all the arcs on the graph of constraints are arc consistent.

```
Function Revise (i, j): Boolean
Begin
   1. CHANGE:= False
   2. For each x ∈ D_i do
   3.   If there is no y ∈ D_j such that R_ij(x, y) then
   4.   Begin
   5.     Delete x from D_i
   6.     CHANGE: = True
   7.     End if
   8. End for
   9. Return CHANGE
End
```

Algorithm 2.1. *The Revise function*

We will review, in what follows, the main AC-i from the related state of the art. To better understand how they work, we propose adopting a simple CSP example throughout this chapter: a map-coloring problem consisting of three variables X_1, X_2 and X_3 with respective domains D_1 = {R, G, B}, D_2 = {R, G} and D_3 = {G}. It is a matter of coloring the vertices of a graph without two adjacent vertices having the same color. The graph of constraints associated with this CSP is illustrated in Figure 2.1.

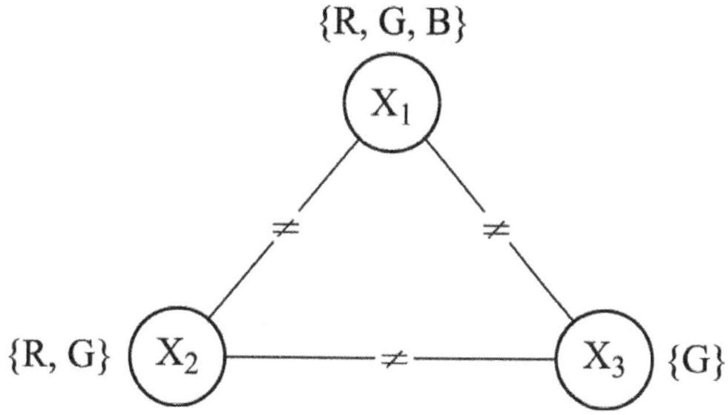

Figure 2.1. *The graph of constraints associated with the map-coloring example*

NOTE:–For the sake of simplicity, we denote by "i" the variable "X_i" in the execution tracks of the following AC-i.

2.2.1. *AC-1*

This early version is due to [ROS 76]. The main mechanism for implementing this procedure is based on a list Q supplied by all the couples of variables (X_i, X_j) and (X_j, X_i) linked by a constraint C_{ij}. The algorithm visits each couple (X_i, X_j) and removes all the values that violate C_{ij} from the domain D_i.

Procedure AC1
Begin

```
1. Q:= {(i, j) | C_ij ∈ C, i≠j}
2. Repeat
3.     CHANGE: = False
4.     For each (i, j) ∈ Q do
5.         CHANGE: = (Revise (i, j) or CHANGE)
6.     End for
7. Until ¬ Change
```

End

Algorithm 2.2. *The AC-1 procedure*

The application of AC-1 to our example is illustrated by the following steps where:

– In the first column, the current couple of variables (i, j) being treated by the *revise* procedure is colored in gray shade.

– In the second column, the value (color) removed by the *revise* procedure is colored in gray shade.

– In the second column, the final domains obtained after performing the whole AC-1 are emphasized in bold.

Iteration	Q	D_i	Change
1.1	{(1, 2); (2, 1); (1, 3); (3, 1); (2, 3); (3, 2)}	D_1 = {R, G, B} D_2 = {R, G} D_3 = {G}	FALSE
1.2	{(1, 2); (2, 1); (1, 3); (3, 1); (2, 3); (3, 2)}	D_1 = {R, G, B} D_2 = {R, G} D_3 = {G}	FALSE
1.3	{(1, 2); (2, 1); (1, 3); (3, 1); (2, 3); (3, 2)}	D_1 = {R, G, B} D_2 = {R, G} D_3 = {G}	TRUE
1.4	{(1, 2); (2, 1); (1, 3); (3, 1); (2, 3); (3, 2)}	D_1 = {R, B} D_2 = {R, G} D_3 = {G}	TRUE

1.5	{(1, 2); (2, 1); (1, 3); (3, 1); (2, 3); (3, 2)}	D_1 = {R, B} D_2 = {R, G} D_3 = {G}	TRUE
1.6	{(1, 2); (2, 1); (1, 3); (3, 1); (2, 3); (3, 2)}	D_1 = {R, B} D_2 = {R} D_3 = {G}	TRUE
2.1	{(1, 2); (2, 1); (1, 3); (3, 1); (2, 3); (3, 2)}	D_1 = {R, B} D_2 = {R} D_3 = {G}	TRUE
2.2	{(1, 2); (2, 1); (1, 3); (3, 1); (2, 3); (3, 2)}	D_1 = {B} D_2 = {R} D_3 = {G}	TRUE
2.3	{(1, 2); (2, 1); (1, 3); (3, 1); (2, 3); (3, 2)}	D_1 = {B} D_2 = {R} D_3 = {G}	TRUE
2.4	{(1, 2); (2, 1); (1, 3); (3, 1); (2, 3); (3, 2)}	D_1 = {B} D_2 = {R} D_3 = {G}	TRUE
2.5	{(1, 2); (2, 1); (1, 3); (3, 1); (2, 3); (3, 2)}	D_1 = {B} D_2 = {R} D_3 = {G}	TRUE
2.6	{(1, 2); (2, 1); (1, 3); (3, 1); (2, 3); (3, 2)}	D_1 = {B} D_2 = {R} D_3 = {G}	TRUE
3.1	{(1, 2); (2, 1); (1, 3); (3, 1); (2, 3); (3, 2)}	D_1 = {B} D_2 = {R} D_3 = {G}	FALSE
3.2	{(1, 2); (2, 1); (1, 3); (3, 1); (2, 3); (3, 2)}	D_1 = {B} D_2 = {R} D_3 = {G}	FALSE
3.3	{(1, 2); (2, 1); (1, 3); (3, 1); (2, 3); (3, 2)}	D_1 = {B} D_2 = {R} D_3 = {G}	FALSE
3.4	{(1, 2); (2, 1); (1, 3); (3, 1); (2, 3); (3, 2)}	D_1 = {B} D_2 = {R} D_3 = {G}	FALSE
3.5	{(1, 2); (2, 1); (1, 3); (3, 1); (2, 3); (3, 2)}	D_1 = {B} D_2 = {R} D_3 = {G}	FALSE
3.6	{(1, 2); (2, 1); (1, 3); (3, 1); (2, 3); (3, 2)}	$\boldsymbol{D_1}$ **= {B}** $\boldsymbol{D_2}$ **= {R}** $\boldsymbol{D_3}$ **= {G}**	FALSE

The systematic nature of the revisions of all the problem's arcs, each time a value is removed, deteriorates the AC-1 performances, since the removal of a value from a variable X_i only has an immediate bearing on the values of the variables that are connected to X_i by a constraint. Thus, for a problem with n variables and d as the maximum size of the domains, the number of constraints is, in the worst case, $n(n - 1)/2$ arcs. The temporal complexity of AC-1 is in $O(n^3d^3)$, while its spatial complexity is in $O(n^2)$.

2.2.2. *AC-2*

To overcome the disadvantages of AC-1, a new version AC-2 was proposed in [MAC 77]. The idea of this algorithm results from the fact that arc consistency can be obtained by testing the neighboring area that consists of the set of variables connected by a binary constraint. In other words, if the revision of the arc led to a change in the domain D_i of the variable X_i then only the arcs (k, i) with $k \leq i$ and $k \neq j$ must be revised.

AC-2 uses two queue structures Q and Q' for storing the arcs to be revised with each iteration. But some arcs may already exist in the data structure and should not then be reincorporated into the queue. This algorithm is a particular case of AC-3 that we will discuss in the following section.

```
Procedure AC-2
Begin
    For i:=1 until n do
    Begin
        Q:= {(i, j) | C_ij ∈ C, I < j}
        Q':= {(j, i) | C_ji ∈ C, j < i}
        While Q not empty do
    Begin
            While Q not empty do
    Begin
                Pop (i, j) from Q
                If Revise (i, j) then
                    Q':=Q' ∪ {(k, i) | C_ki ∈ C, k ≤ i, k ≠ j}
                End if
            End while
            Q:= Q'
            Q':= Ø
        End while
    End for
End
```

Algorithm 2.3. *The AC-2 procedure*

By applying AC-2 to our example, we obtain the following steps:

Iteration	X_i	Q	Q′	D_i
1.1	1	{(1, 2); (1, 3)}	∅	D_1 = {R, G, B} D_2 = {R, G} D_3 = {G}
1.2	1	{(1, 3)} ∅	∅	D_1 = {R, G, B} D_2 = {R, G} D_3 = {G}
2.1	2	{(2, 3)} ∅	∅ {(1, 2)}	D_1 = {R, B} D_2 = {R, G} D_3 = {G}
2.2	2	{(1, 2)} ∅	∅	D_1 = {R, B} D_2 = {R} D_3 = {G}
3.1	3	{(3, 1); (3, 2)}	∅	D_1 = {B} D_2 = {R} D_3 = {G}
3.2	3	{(3, 2)} ∅	∅	**D_1 = {B}** **D_2 = {R}** **D_3 = {G}**

The temporal complexity of AC-1 is around $O(n^2d^3)$.

2.2.3. *AC-3*

The version AC-3 [MAC 77], just like the version AC-2, is a clear improvement on the early version of AC-1. It uses the same property as AC-2 to limit the number of arc revisions inferred by each removal of a value. This algorithm uses a single queue structure. To determine all non-viable values, AC-3 seeks a support for each value on each constraint. For this, it uses a list Q of pairs of the variables to be studied. This structure is then traversed and each arc (i, j) is then revised using the same *Revise* procedure. During this revision, if the removal of value v_i occurs in D_i, the set of arcs (k, i) such as $k \neq i$ and $k \neq j$ are added to Q (if the structure does not already contain them). AC-3 then re-examines the viability of all the values v_k of D_k in relation to C_{ki}. It may well be that (i, v_i) is the sole support for certain values of D_k and that the removal of v_i makes them not arc consistent. This algorithm re-examines the viability of more values than necessary (re-examines all the values even those that are not concerned by the removal). Therefore, for a complete graph of constraints, the structure F will see the insertion of $O(n^2d)$ arcs hence the temporal complexity $O(n^2d^3)$ and the spatial complexity $O(n^2)$ of AC-3.

Compared to AC-1, AC-3 is more efficient; its complexity in the worst case is in $O(md^3)$.

Procedure AC3
Begin

1. Q:= $\{(i, j) \mid C_{ij} \in C\ i \neq j\}$
2. While Q not empty Do
3. Begin
4. Select or delete a pair (i, j) from Q
5. If Revise (i, j) then
6. Q:= Q $\cup$ $\{(k, i) \mid C_{ki} \in C, k \neq i \text{ et } k \neq j\}$
7. End if
8. End while

End

Algorithm 2.4. *The AC-3 procedure*

The following phases illustrate the application of AC-3 to our map-coloring example:

Iteration	**Q**	**D$_i$**	**Revise (i, j)**
1	{(1, 2); (2, 1); (1, 3); (3, 1); (2, 3); (3, 2)}	D1 = {R, G, B} D2 = {R, G} D3 = {G}	FALSE
2	{(2, 1); (1, 3); (3, 1); (2, 3); (3, 2)}	D1 = {R, G, B} D2 = {R, G} D3 = {G}	FALSE
3	{(1, 3); (3, 1); (2, 3); (3, 2)}	D1 = { R, G, B} D2 = {R, G} D3 = {G}	TRUE
4	{(3, 1); (2, 3); (3, 2) ∪ **(2, 1)**}	D1 = {R, B} D2 = {R, G} D3 = {G}	FALSE
5	{(2, 3); (3, 2); (2, 1)}	D1 = {R, B} D2 = {R, G} D3 = {G}	TRUE
6	{(3, 2); (2, 1) ∪ **(1, 2)**}	D1 = {R, B} D2 = {R} D3 = {G}	FALSE
7	{(2, 1); (1, 2)}	D1 = {R, B} D2 = {R} D3 = {G}	FALSE
8	{(1, 2)}	D1 = {R, B} D2 = {R} D3 = {G}	TRUE
9	{∅ ∪ **(3, 1)**}	D1 = {B} D2 = {R} D3 = {G}	FALSE
10	{∅}	**D1 = {B}** **D2 = {R}** **D3 = {G}**	FALSE

2.2.4. *AC-4*

This method constitutes the first version of AC-i, which has an optimal complexity in the worst case [MOH 86]. Unlike previous versions, the fourth version no longer relies on the *Revise* procedure. It takes place in two successive phases:

2.2.4.1. *Phase 1: initialization*

In this first phase, the algorithm AC-4 determines:

– for any couple of variables (X_i, X_j) connected by a constraint C_{ij} and for any element a of D_i, the number of elements of the domain D_j that are compatible with a, denoted by *cpt* $[(i, j), a]$ and commonly known as support of a;

– for any element b of D_j, the *Support set* S_{jb} of the couples (i, a) such as X_i and X_j are connected by a constraint C_{ij} and a is compatible with b.

Each value $a \in D_i$ having a *cpt* $[(i, j), a]$ equal to zero (the value a of X_i has no support in Dj) must be removed from D_i and placed in a waiting list L to propagate its removal in the second propagation phase, then a Boolean variable *mark* $[i, a]$ is set to 1.

2.2.4.2. *Phase 2: propagation*

The information established during the first phase is preserved. Indeed, the removal of the marked values with a zero counter *cpt* $[(i, j), a]$ causes the decrementation of the counters of all the values still present, which were compatible with a and which were referenced in *Support* S_{ia}. This phase can be repeated when new counters reach the value of 0.

The complexity of this algorithm is linear in its size since during propagation an edge is only visited once in one direction. If m is the number of constraints and d the maximum size of the domains, the temporal complexity of

AC-4 can be bounded from above by $O(md^2)$ [JÉG 91]. On the other hand, AC-5, which is a generic version, is of the complexity in $O(md)$ in some cases [VAN 92].

```
Procedure AC4
Begin
//Phase 1: Initialization
    Mark:= 0; S_ia:= Ø
    For each (i, j) | C_ij ∈ C do
        For each a ∈ D_i do
            Total:= 0
            For each b ∈ D_j do
                If R (i, a, j, b) then
            /*R returns True if a and b are compatible*/
                    Total:= Total + 1
                    S_jb:= S_jb ∪ {(i, a)}
            End if
        End for
        If total = 0 then
            Mark [i, a] = 1
            D_i:= D_i \ {a}
        Else
            cpt [(i, j), a)]:= Total
        End if
    End for
End for
    Initialize L with {(i, a) | mark [i, a] = 1}
```

```
// Phase 2: Propagation
    While L not empty do
        Select and delete (j, b) from L
          For each (i, a) ∈ S_jb do
              cpt [(i, j), a]:= cpt [(i, j), a]-1
          If cpt [(i, j), a] = 0 and mark [i, a] = 0 then
                L:= L ∪ {(i, a)}
                Mark [i, a]:= 1
                D_i:= D_i \ {a}
            End if
          End for
    End while
End
```

Algorithm 2.5. *The AC-4 procedure*

The application of AC-4 to our example is illustrated by the following steps:

Phase	*(i,j)*	*cpt* [*(i,j), a*]/*mark* [*i, a*]	*Sjb*
INITIALIZATION	(1, 2)	*cpt* [(1, 2), R] = 1	S_{2G}={(1, R)}
		cpt [(1, 2), G] = 1	S_{2R}={(1, G)}
		cpt [(1, 2), B] = 2	S_{2G}={(1, R); (1, B)} S_{2R}={(1, G); (1, B)}
	(2, 1)	*cpt* [(2, 1), R] = 2	S_{1G} = {(2, R)} S_{1B} = {(2, R)}
		cpt [(2, 1), G] = 2	S_{1R} = {(2, G)} S_{1B} = {(2, R); (2, G)}
	(1, 3)	*cpt* [(1, 3), R] = 1	S_{3G}= {(1,R)}
		mark [1, G] = 1	$D_1 = D_1 - \{G\}$
		cpt [(1, 3), B] = 1	S_{3G} = {(1, R); (1, B)}

	(3, 1)	*cpt* [(3, 1), G] = 2	S_{1R} = {(2, G); (3, G)} S_{1B} = {(2, R); (2, G); (3, G)}
	(2, 3)	*cpt* [(2, 3), R] = 1	S_{3G} = {(1, R); (1, B); (2, R)}
		mark [2, G] = 1	$D_2 = D_2$ – {G}
	(3, 2)	*cpt* [(3, 2), G] = 1	S_{2R} = {(1, G); (1, B); (3, G)}
D_1 = {R, B}; D_2 = {R}; D_3 = {G}; L = {(1, G); (2, G)}			
PROPAGATION	(1, G)	*cpt* [(2, 1), R] = 1	S_{1G} = {(2, R)}
	(2, G)	*cpt* [(1, 2), R] = 0 *mark* [1, R] = 1	S_{2G} = {(1, R); (1, B)} $D_1 = D_1$ – {R} L = L ∪ {(1, R)}
		cpt [(1, 2), B] = 1	S_{2G} = {(1, R); (1, B)}
	(1, R)	*cpt* [(2, 1), G] = 1	S_{1R} = {(2, G); (3, G)}
		cpt [(3, 1), G] = 1	S_{1R} = {(2, G); (3, G)}
D_1 = {B}; D_2 = {R}; D_3 = {G}			

2.2.5. *AC-5*

AC-5 [VAN 92] is a generic arc-consistency algorithm that can be reduced to AC-3 or AC-4. It may be adapted to several classes of constraints, such as functional, anti-functional and monotonic constraints.

The queue in AC-5 contains elements of the form $[(i, j), w]$ such that (i, j) is an arc in the graph of constraints and w is the value removed from Dj, which justified the need to revise the arc (i, j). In fact, the queue is organized so as to contain elements of the form $[\{A_1, ..., A_m\}, v]$ such that A_k is an arc and v is a value with $m > 0$.

Procedure **INITQUEUE** (out Q)

Post: $Q = \varnothing$

Algorithm 2.6. *The INITQUEUE procedure*

Algorithm 2.6 initializes the queue to the empty set.

Function **EMPTYQUEUE** (in Q): Boolean

Post: EMPTYQUEUE $\leftrightarrow (Q = \varnothing)$

Algorithm 2.7. *The EMPTYQUEUE procedure*

Algorithm 2.7 tests whether the queue is empty or not.

Procedure DEQUEUE (in out Q, out I, j, w)

Post: $[(i, j), w] \in Q_0$ and $Q = Q_0 \setminus [(i, j), w]$

Algorithm 2.8. *The DEQUEUE procedure*

Algorithm 2.8 consists of removing an element from the queue. It takes an element $[\{A_1, ..., A_m\}, w]$ and removes an arc A_k in the form of (i, j) and returns i, j and w.

Procedure ENQUEUE (in i, Δ, in out Q)

Pre: $\Delta \subseteq D_i$

Post: $Q = Q_0 \cup \{[(k,i), v] \mid (k, i)$ is an arc and $v \in \Delta\}$

Algorithm 2.9. *The ENQUEUE procedure*

This procedure is used each time that the subset of values Δ is removed from D_i. It introduces elements in the form of $[(k, i), v]$ into the queue such that (k, i) is an arc of the graph of constraints and $v \in \Delta$.

The ENQUEUE procedure (i, Δ, Q) adds an element $[\{A_1, \ldots, A_m\}, v]$ to the queue such that A_k is an arc in the form of (j, i) for each value $v \in \Delta$.

Procedure ARCCONS (in i, j, out Δ)

Pre: (i, j) is an arc, $D_i \neq \emptyset$ and $D_j \neq \emptyset$

Post: $\Delta = \{v \in D_i \mid \forall w \in D_j: \neg C_{ij}(v, w)\}$

Algorithm 2.10. *The ARCCONS procedure*

The *ARCCONS procedure* (i, j, Δ) calculates all the values of Δ for the variable X_i that are not supported by D_j.

Procedure LOCALARCCONS (in i, j, w, out Δ)

Pre: (i, j) is an arc, $w \notin D_j$, $D_i \neq \emptyset$ and $D_j \neq \emptyset$

Post: $\Delta_{1=} \{v \in D_i \mid C_{ij}(v, w \text{ and } \mid \forall w' \in D_j: \neg C_{ij}(v, w')\}$

$\Delta_{1=} \{v \in D_i \mid \forall w' \in D_j: \neg C_{ij}(v, w')\}$

Algorithm 2.11. *The LOCALARCCONS procedure*

The *LOCALARCCONS procedure* (i, j, w, Δ) is used to calculate all of the values in $\boldsymbol{D}_i$ that no longer have support due to the removal of the value w from $\boldsymbol{D}_j$.

This *LOCALARCCONS* specification gives us more room for the result to return Δ. It is therefore important to calculate Δ_1 to ensure the accuracy of AC-5 and calculate Δ_2 to prevent overpruning.

In fact, the AC-5 algorithm consists of two steps. First, all arcs are considered once and the arc consistency is reinforced on each one of them. The REMOVE (Δ, D_i) procedure consists of removing the subset Δ from the domain of the variable X_i. The second step consists of applying the LOCALARCCONS procedure for each element of the queue with the possibility of inserting new elements.

```
Procedure AC5
Begin
    INITQUEUE (Q)
    For each (i, j) | (i, j) is an arc do
        ARCCONS (i, j, Δ)
        ENQUEUE (i, Δ, Q)
        REMOVE (Δ, D_i)
    End for
    While (¬ EMPTYQUEUE (Q)) do
        DEQUEUE (Q, i, j, w)
        LOCALARCCONS (i, j, w, Δ)
        ENQUEUE (i, Δ, Q)
        REMOVE (Δ, D_i)
    End while
End
```

Algorithm 2.12. *The AC-5 procedure*

In the following, we present the track of the AC-5 procedure on our initial example:

Phase	(i, j)	
PHASE I	(1, 2)	ARCCONS (1, 2, Δ) ➔ Δ = { } ENQUEUE (1, Δ, Q) ➔ Q= Q ∪ { } REMOVE (Δ, D_1) ➔ D_1 = D_1 – { }
	(2, 1)	ARCCONS (2, 1, Δ) ➔ Δ = { } ENQUEUE (2, Δ, Q) ➔ Q= Q ∪ { } REMOVE (Δ, D_2) ➔ D_2 = D_2 – { }
	(1, 3)	ARCCONS (1, 3, Δ) ➔ Δ = {G} ENQUEUE (1, Δ, Q) ➔ Q= Q ∪ {< (3, 1), G>, < (2, 1), G>} REMOVE (Δ, D_1) ➔ D_1 = D_1 – {G} = {R, B}
	(3, 1)	ARCCONS (3, 1, Δ) ➔ Δ = { } ENQUEUE (3, Δ, Q) ➔ Q= Q ∪ { } REMOVE (Δ, D_3) ➔ D_3 = D_3 – { }
	(2, 3)	ARCCONS (2, 3, Δ) ➔ Δ = {G} ENQUEUE (2, Δ, Q) ➔ Q= Q ∪ {< (3, 2), G>, < (1, 2), G>} REMOVE (Δ, D_2) ➔ D_2 = D_2 – {G} = {R}
	(3, 2)	ARCCONS (3, 2, Δ) ➔ Δ = { } ENQUEUE (3, Δ, Q) ➔ Q = Q ∪ { } REMOVE (Δ, D_3) ➔ D_3 = D_3 – { }
D_1 = {R, B}; D_2 = {R}; D_3 = {G}; Q= {< (3, 1), G>, < (2, 1), G>, < (3, 2), G>, < (1, 2), G>}		
PHASE II	<(3, 1), G>	DEQUEUE (Q, 3, 1, G) ➔ Q = Q – {< (3, 1), G>} LOCALARCCONS (3, 1, G, Δ) ➔ Δ_1= { } Δ_2= { } Δ= { } ENQUEUE (3, Δ, Q) ➔ Q= Q ∪ { } REMOVE (Δ, D_3) ➔ D_3 = D_3 – { }

	<(2, 1), G>	DEQUEUE (Q, 2, 1, G) ➔ Q = Q – {< (2, 1), G>} LOCALARCCONS (2, 1, G, Δ) ➔ Δ_1= { } Δ_2= { } Δ= { } ENQUEUE (2, Δ, Q) ➔ Q= Q ∪ { } REMOVE (Δ, D_2) ➔ $D_2 = D_2$ – { }
	<(3, 2), G>	DEQUEUE (Q, 3, 2, G) ➔ Q = Q – {< (3, 2), G>} LOCALARCCONS (3, 2, G, Δ) ➔ Δ_1= { } Δ_2= { } Δ = { } ENQUEUE (3, Δ, Q) ➔ Q= Q ∪ { } REMOVE (Δ, D_3) ➔ $D_3 = D_3$ – { }
	<(1, 2), G>	DEQUEUE (Q, 1, 2, G) ➔ Q = Q – {< (1, 2), G>} LOCALARCCONS (1, 2, G, Δ) ➔ Δ_1= {R} Δ_2= {R} Δ = {R} ENQUEUE (1, Δ, Q) ➔ Q= Q ∪ {< (2, 1), R>, < (3, 1), R>} REMOVE (Δ, D_1) ➔ $D_1 = D_1$ – {R} = {B}
	<(2, 1), R>	DEQUEUE (Q, 2, 1, R) ➔ Q = Q – {< (2, 1), R>} LOCALARCCONS (2, 1, R, Δ) ➔ Δ_1= { } Δ_2= { } Δ = { } ENQUEUE (2, Δ, Q) ➔ Q= Q ∪ {} REMOVE (Δ, D_2) ➔ $D_2 = D_2$ – { }
	<(3, 1), R>	DEQUEUE (Q, 3, 1, R) ➔ Q = Q – {< (3, 1), R>} LOCALARCCONS (3, 1, R, Δ) ➔ Δ_1= { } Δ_2= { } Δ = { } ENQUEUE (3, Δ, Q) ➔ Q= Q ∪ { } REMOVE (Δ, D_3) ➔ $D_3 = D_3$ – { }
D_1 = {B}; D_2 = {R}; D_3 = {G}		

The spatial complexity of AC-5 is around $O(md)$ while its temporal complexity depends on the complexity of the two *ARCCONS* and *LOCALARCCONS* procedures: $O(md^2)$ and $O(md)$.

AC-5's contribution then occurs at the level of the queue structure and at the level of complexity. Indeed, the advantage of AC-5 compared with AC-3 and AC-4 is the use of constraint semantics to reduce the temporal complexity down to $O(md)$ [VAN 92].

2.2.6. *AC-6*

```
Procedure AC6
Begin
//Phase 1: Initialization
L:= Ø ;
For each (i, a) ∈ D_i do
    If a:= Ø;
    Mark (i, a):= 0
End for
    For each (i, j) | C_ij ∈ C do
        For a ∈ D_i do
          If D_j:= Ø then
          Empty support = True
          Else
               b: = first (D_j)
               NextSupport (i, j, a, b, EmptySupport)
               If EmptySupport then
                           D_i:= D_i \ {a}
                           Mark (i, a):= 1
                           L:= L ∪ {(i, a)}
              Else
                           S_jb:= S_jb ∪ {(i, a)}
             End if
        End if
      End for
  End for
```

The aim of AC-6 proposed in [BES 93] is to avoid the costly constraint tests carried out by AC-4, to find all the supports for all the values. Indeed, the arc consistency requires a single support for the variable on each one of the constraints linking this variable to other variables. It is therefore not necessary to know all the supports of a value or even how many. AC-6 maintains the same principle as AC-4. But, instead of looking for all the supports of a value, it only looks for one support (the first) for each (i, a) in Dj on each constraint C_{ij} to prove that (i, a) is viable, i.e. it can contribute to a solution. To allow for incremental computation of supports, an order relation is introduced into each domain.

The principle of this algorithm consists of looking for the first support of each value a in relation to each arc (i, j). When a value b in a domain Dj is found, (i, a) is added to the list *Support* S_{jb}. If the value b is removed from Dj, this leads to the search for a new support c using b for each value (i, a) belonging to *Support* S_{jb} in Dj.

Thus, the spatial complexity of AC-6 is, in the worst case, in $O(n^2d^2)$ while its temporal complexity is equal to that of AC-4 that is in $O(n^2d^2)$.

```
//Phase 2: Propagation
    While L ≠ Ø do
        Select and delete (j, b) from L
        For each (i, a) ∈ S_jb do
            S_jb:= S_jb \ {(i, a)}
            If mark (i, a) = 0 then
                c:= b
                NextSupport (i, j, a, c, EmptySupport)
                If EmptySupport then
                    D_i:= D_i \ {a}
                    Mark (i, a) := 1
                    L:= L ∪ {(i, a)}
            Else
                S_jc:= S_jc ∪ {(i, a)}
            End if
        End if
      End for
    End while
End
```

Algorithm 2.13. *The AC-6 procedure*

```
Procedure NextSupport (i, j, a, b, EmptySupport)
Begin
    If b ≤ last (D_j) then
        EmptySupport:= False
        While mark (j, b) = 1 do
            b:= b+1
        End while
        While ¬ R_ij (a, b) and ¬ EmptySupport do
            If b < last (D_j) then
                b:= next (b, D_j)
            Else
                EmptySupport:= True
            End if
        End while
    Else
        EmptySupport:= True
    End if
End
```

Algorithm 2.14. *The NextSupport procedure*

In what follows, we give the track of the AC-6 algorithm applied to our example:

Phase	*(i,j)*	[(i, j), a]/*mark* [i, a]	EmptySupport	*Sjb*
INITIALIZATION	(1, 2)	[(1, 2), R]	FALSE	S_{2G} = {(1, R)}
		[(1, 2), G]	FALSE	S_{2R} = {(1, G)}
		[(1, 2), B]	FALSE	S_{2R} = {(1, G); (1, B)}
	(2, 1)	[(2, 1), R]	FALSE	S_{1G} = {(2, R)}
		[(2, 1), G]	FALSE	S_{1R} = {(2, G)}
	(1, 3)	[(1, 3), R]	FALSE	S_{3G} = {(1, R)}
		mark [1, G] = 1	TRUE	$D_1 = D_1 - \{G\}$
		[(1, 3), B]	FALSE	S_{3G} = {(1, R); (1, B)}
	(3, 1)	[(3, 1), G]	FALSE	S_{1R} = {(2, G); (3, G)}

	(2, 3)	[(2, 3), R]	FALSE	S_{3G} = {(1, R); (1, B); (2,R)}
		mark [2, G] = 1	TRUE	$D_2 = D_2 - \{G\}$
	(3, 2)	[(3, 2), G]	FALSE	S_{2R} = {(1, G); (1, B); (3, G)}
D_1 = {R, B}; D_2 = {R}; D_3 = {G}; *L* = {(1, G); (2, G)}				
PROPAGATION	(1, G)	[(2, 1), R]	FALSE	S_{1G} = { } S_{1B}={(2, R)}
	(2, G)	*mark* [1, R] = 1	TRUE	S_{2G} = { } $D_1 = D_1 - \{R\}$; L=L∪{(1,R)}
	(1, R)	[(2, 1), G]	–	S_{1R} = {(3, G)}
		[(3, 1), G]	FALSE	S_{1R}={ } S_{1B}={(2,R); (3, G)}
		D_1 = {B}; D_2 = {R}; D_3 = {G}		

There is another algorithm denoted by AC6++ [BES 95], called AC inference, which consists, as its name suggests, of inferring new knowledge from those already established. The inference process is principally based on the notion of bidirectional support: let us assume that the consistency between (i, a) and (j, b) is verified, we obtain that (j, b) supports (i, a), i.e. that b from Dj is compatible with a. We then infer that (i, a) supports (j, b). In addition to this property, there are other notions that allow this process to be activated. Indeed, from a tuple (a, b) authorized by the constraint C_{ij}, the commutativity can infer that (b, a) is in turn authorized by the constraint C_{ij}. As for irreflexivity, it can be used, when (a, a) is a forbidden tuple for the constraint C_{ij}, to infer that (i, a) is inconsistent with (j, a).

2.2.7. *AC-7*

The AC-7 algorithm constitutes a general arc-consistency procedure [BES 95]. It brings refinements to the AC-6 procedure. It preserves the basic idea of AC-6 while avoiding a large number of consistency tests. This algorithm is optimal in terms of the number of consistency tests carried out. It uses the fact that the constraints are not directed. This is in fact due to the use of the bidirectionality property of the relations associated with the constraints.

Algorithms 2.15–2.17 show how AC-7 works. Before detailing AC-7, we recall the definition of bidirectionality.

```
Procedure AC7
Begin
    SeekSupportSet:= Ø
    For each (i, j) | C_ij ∈ C do
        For each a ∈ D_i do
            CS [i, a, j]:= Ø
            Last [i, a, j]:= 0 // The first value of D_j minus 1
            Push [(i, a), j] in SeekSupportSet
        End for
    End for
    While SeekSupportSet ≠ Ø do
        Pop [(i, a), j] from SeekSupportSet
        If a ∈ D_i then
            c:= SeekCurrentSupport (i, a, j)
            If c ≠ nul then
                Push a in CS [j, c, i]
            Else
                ProcessDeletion (i, a, SeekSupportSet)
            End if
        End if
    End while
End
```

Algorithm 2.15. *The AC-7 procedure*

DEFINITION 2.5:– BIDIRECTIONAL CONSTRAINT.– *A constraint is called bidirectional if the compatibility of* (i, a) *with* (j, b) *results in the compatibility of* (j, b) *with* (i, a). *All the constraints verify this property, and consequently*:

$$\forall\, C_{ij} \in C, \forall\, (a, b) \in D_i \times D_j;\ C_{ij}\,(a, b) = C_{ji}\,(b, a).$$

Given that bidirectionality is a general property of constraints, it can be used to reduce the number of constraint checks. AC-7 uses the same data structure as AC-6 and is based on the following properties:

– Never check $R_{ij}(a, b)$ if there is a value $b' \in D_j$ such that $R_{ij}(a, b')$ has already been proved to be consistent.

– Never check $R_{ij}(a, b)$ if there is a value $b' \in D_j$ such that $R_{ji}(b', a)$ has already been proved to be consistent.

– Never check $R_{ij}(a, b)$ if it has already been checked or if $R_{ji}(b, a)$ has been checked.

This algorithm contains two main functions: the first "*SeekCurrentSupport*", which enables a current support for a given value to be sought, and the second "*ProcessDeletion*", which enables the removal of a value to be propagated.

Initially for each value (i, a), on each constraint C_{ij}, the value–variable couple $[(i, a), j]$ is inserted into the SeekSupportSet list. For each value–variable couple $[(i, a), j]$ in SeekSupportSet, AC-7 seeks a support for (i, a) in D_j. This is done by using the *SeekCurrentSupport* function. If a value c is found, then a is added to CS $[j, c, i]$ since c now supports (i, a) for C_{ij}.

During the search, if a value (i, a) is found without support for a given constraint, it is removed from D_i and this removal is handled by adding all the value–variable couples $[(j, b), i]$ to the SeekSupportSet list knowing that (j, b) was supported by (i, a) (i.e. the values (j, b) that were in the set CS $[i, a, j]$).

When SeekCurrentSupport is used to find a support for (i, a) in D_j, it tests whether CS $[i, a, j]$ still contains values from D_j or not. Indeed, if the list contains values from D_j, then searching for a support for (i, a) is useless since (i, a) already has inferable supports (the values supported by (i, a) are also support values for (i, a)).

Each time that SeekCurrentSupport finds a value b in CS $[i, a, j]$ that does not belong to D_j, b is removed from CS $[i, a, j]$ to avoid its reverification.

If no inferable support is found, then the search for a new support for (i, a) is launched by seeking the smallest value in D_j that is greater than or equal to Last $[i, a, j]$ and supporting (i, a) for C_{ij}. This is done by avoiding retesting $R_{ij}(a, b)$ if $R_{ij}(b, a)$ is already proved to be inconsistent (i.e. that we know that (j, b) has no support in D_i up to a value greater than a).

```
Function Process Deletion (i, a, SeekSupportSet)
Begin
    Delete a from D_i
    If D_i = Ø then
    Exit
    End if
    For each j | R_ij ∈ R do
      For each b ∈ CS[i, a, j] do
          Delete b from CS [i, a, j]
          Push [(j, b), i] in SeekSupportSet /* b may be a
              support for another value */
    End for
    End for
End
```

Algorithm 2.16. *The ProcessDeletion function*

```
Function SeekCurrentSupport (i, a, j)//returns a value that
Begin                                   supports (i, a) or nul if ∄
    isFound:= False
    While CS [i, a, j] ≠ ∅ and ¬ isFound do
        b:= one element of CS [i, a, j]
        If b ∈ D_j then
            isFound:= True
        Else
            Delete b from CS [i, a, j]
        End if
    End while
    If isFound then
        Return b
    End if
    b:= Last [i, a, j]
    If b > highest (D_j) then
        Return nul
    End if
    While b ∉ D_j do
        b:= b+1
    End while
    While b ≠ nul and ¬ inFound do
        If Last [j, b, i] ≤ a and (R_ij (a, b)) then
            inFound:= True
        Else
            b:= next (b, D_j)
        End if
    End while
    Last [i, a, j]:= b
    Return b
End
```

Algorithm 2.17. *The SeekCurrentSupport function*

By applying AC-7 to our example, we obtain the following phases:

For each pair of variables (i, j), for each value a from Di, we initialize CS $[i, a, j]$ to the empty set and Last $[i, a, j]$ to zero. Then, we add this value–variable couple to the SeekSupportSet list.

SeekSupportSet = {[(1, R), 2]; [(1, G), 2]; [(1, B), 2]; [(1, R), 3]; [(1, G), 3]; [(1, B), 3]; [(2, R), 1]; [(2, G), 1]; [(2, R), 3]; [(2, G), 3]; [(3, G), 1]; [(3, G), 2]}

– [(1, R), 2]

SeekSupportSet ≠ ∅; CS [1, R, 2] =∅; Last [1, R, 2] = 0;

b: = Last [1, R, 2] = 0 $\notin D_2$; b:= b + 1 = R;

Last [2, R, 1] = 0 < = Last [1, R, 2] but C_{12} *is violated*;

therefore: b:= b + 1 = G;

Last [2, G, 1] = 0 < = R and C_{12} *is satisfied*;

therefore: Last [1, R, 2] = G; c:= G; CS [2, G, 1]:=R *(by inference)*.

– [(1, G), 2]

SeekSupportSet ≠ ∅; CS [1, G, 2] = ∅; Last [1, G, 2] = 0;

b:= Last [1, G, 2] = 0 $\notin D_2$; b:= b + 1 = R;

Last [2, R, 1] = 0 < = G and C_{12} *is satisfied*;

therefore: Last [1, G, 2] = R; c:= R; CS [2, R, 1]:= G *(by inference)*.

– [(1, B), 2]

SeekSupportSet ≠ ∅; CS [1, B, 2] = ∅; Last [1, B, 2] = 0;

b:= Last [1, B, 2] = 0 $\notin D_2$; b := b + 1 = R;

Last [2, R, 1] = 0 < = B and C_{12} *is satisfied*;

therefore: Last [1, B, 2] = R; c := R; CS [2, R, 1]:= B *(by inference)*.

– [(1, R), 3]

SeekSupportSet ≠ ∅; CS [1, R, 3] = ∅; Last [1, R, 3] = 0;

b:= Last [1, R, 3] = 0 $\notin D_3$; b := b + 1 = G;

Last [3, G, 1] = 0 < = R and C_{13} *is satisfied*;

therefore: Last [1, R, 3] = G; c := G; CS [3, G, 1] := R *(by inference)*.

– [(1, G), 3]

SeekSupportSet $\neq \emptyset$; CS [1, G, 3] = $\emptyset$; Last [1, G, 3] = 0;

b := Last [1, G, 3] = 0 $\notin D_3$; b:= b + 1 = G;

Last [3, G, 1] = 0 < = G but C_{13} *is violated;*

therefore: b:= b + 1 = nil; Last [1, G, 3] = nil; c:= nil; D_1 := $D_1 \backslash \{G\}$ = {R, B};

since: CS [1, G, 2] = CS [1, G, 3] = $\emptyset$; therefore we have no rectification.

– [(1, B), 3]

SeekSupportSet $\neq \emptyset$; CS [1, B, 3] = $\emptyset$; Last [1, B, 3] = 0;

b := Last [1, B, 3] = 0 $\notin D_3$; b := b + 1 = G;

Last [3, G, 1] = 0 < = B and C_{13} *is satisfied;*

therefore: Last [1, B, 3] = G; c := G ; CS [3, G, 1] := B *(by inference)*.

– [(2, R), 1]

SeekSupportSet $\neq \emptyset$; CS [2, R, 1] = {G, B}; Last [2, R, 1] = 0;

b := G $\notin D_1$; CS [2, R, 1] := CS [2, R, 1]\{G} = {B}; b := B $\in D_1$;

therefore c := B; CS [1, B, 2] := R *(by inference)*.

– [(2, G), 1]

SeekSupportSet $\neq \emptyset$; CS [2, G, 1] = {R}; Last [2, G, 1] = 0;

b := R $\in D_1$; therefore: c:= R; CS [1, R, 2] := G *(by inference)*.

– [(2, R), 3]

SeekSupportSet $\neq \emptyset$; CS [2, R, 3] = $\emptyset$; Last [2, R, 3] = 0;

b := Last [2, R, 3] = 0 $\notin D_3$; b := b + 1 = G;

Last [3, G, 2] = 0 < = R and C_{23} *is satisfied;*

therefore: Last [2, R, 3] = G; c := G; CS [3, G, 2] := R *(by inference).*

– [(2, G), 3]

SeekSupportSet ≠ ∅; CS [2, G, 3] = ∅; Last [2, G, 3] = 0;

b : = Last [2, G, 3] = 0 ∉ D_3; b := b + 1 = G;

Last [3, G, 2] = 0 < = G but C_{23} *is violated*;

therefore: b : = b + 1 = nil; Last [2, G, 3] = nil; c := nil; D_2 := D_2\{G} = {R};

since: CS [2, G, 1] = {R}; CS [2, G, 1] = CS [2, G, 1]\{R} = ∅;

SeekSupportSet := SeekSupportSet ∪ [(1, R), 2] and CS [2, G, 3] = ∅

therefore, we have no rectification.

SeekSupportSet = {[(3, G), 1]; [(3, G), 2]; [(1, R), 2]}.

– [(3, G), 1]

SeekSupportSet ≠ ∅; CS [3, G, 1] = {R, B}; Last [3, G, 1] = 0;

b := R ∈ D_1; therefore: c:= R; CS [1, R, 3] := G *(by inference).*

– [(3, G), 2]

SeekSupportSet ≠ ∅; CS [3, G, 2] = {R}; Last [3, G, 2] = 0.

b := R ∈ D_2; therefore: c := R; CS [2, R, 3] := G *(by inference).*

– [(1, R), 2]

SeekSupportSet ≠ ∅; CS [1, R, 2] = {G}; Last [1, R, 2] = G;

b := G ∉ D_2; CS [1, R, 2] = CS [1, R, 2]\{G} = ∅;

b := Last [1, R, 2] = G; b > highest (D_2) = R;

therefore: c := nil; D_1 := D_1\{R} = {B};

since: CS [1, R, 2] = CS [1, R, 3] = ∅;

therefore, we have no rectification.

We have SeekSupportSet = ∅; D_1 = {B}; D_2 = {R}; D_3 = {G}

AC-7 has, in the worst case, the same complexities as AC-6. However, this algorithm carries out fewer consistency tests, which is no small matter when these tests are so expensive.

The spatial complexity of AC-7 is in $O(md)$, while its temporal complexity is in $O(md^2)$.

The different arc-consistency reinforcement algorithms detailed in this section generally receive inconsistent values for a variable and must propagate these inconsistencies through removals. They are based on two types of constraint propagation scheme:

– Constraint-oriented propagation scheme (AC-1, AC-2 and AC-3);

– Value-oriented propagation scheme (AC-4, AC-6 and AC-7).

2.2.8. *AC2000*

AC-3 is the most frequently used generic algorithm and the easiest to implement since it can be based on a constraint or even on a variable-oriented scheme. [BES 01] has proposed a new algorithm AC2000, which is an improvement of AC-3.

If we examine the behavior of AC-3, we will see that the removal of a single value b from the domain D_j of the variable X_j is sufficient to insert X_j in the queue of the variables to be propagated. Subsequently, any constraint C_{ij} involving X_j is then re-examined. The algorithm then seeks a support for each value from the domain of X_i even though some values of D_i are not supported by the value b removed from D_j. Let us consider the following example. Let the equality constraint be C_{ij} with $D_i = D_j = \{1, ..., 11\}$. The removal of the value 11 from D_j leads to seeking a support for each value of Di. The total cost of seeking (expressed in number of constraint checks or constraint tests performed) is equal to $1 + 2 + ... + 9 + 10 + 11 = 65$ constraint tests, while it is only value 11 of X_i that has lost its support.

[BES 01] proposes not to make a blind research of a support for each value from the domain of X_i each time the domain of X_j is modified, but rather under certain conditions. The idea is to use another data structure denoted by $\Delta(j)$, which, for each variable X_j, contain the values removed following its last propagation. Thus, for each value b removed from D_j, the algorithm first checks if *the value a* of D_i has really lost a support, i.e. whether one of its supports belongs to $\Delta(j)$. The larger the structure $\Delta(j)$, the more costly the process and the higher the probability of finding a support for a in $\Delta(j)$. That is why the authors have used a ratio (threshold) that is equal to the ratio between $\Delta(j)$ and Dj. Verification of the existence of a support in $\Delta(j)$ is only carried out if and only if $|\Delta(i)| < Ratio \times |Dj|$. For our example, $\Delta(j) = \{11\}$. The algorithm then verifies, for each value from the domain of X_i, if it possesses a support in $\Delta(j)$. This procedure requires 11 tests just for value 11 of X_i for which we seek a new support. This requires 10 additional tests. The total number of tests carried out is then 21 instead of 65 tests for AC-3.

The spatial complexity of AC2000 is in $O(nd)$ if this algorithm uses a variable-oriented propagation scheme. This complexity is in $O(nd)$ while using a constraint-oriented propagation scheme. As for its temporal complexity, it is in $O(nd^3)$.

```
Function AC2000(X): Boolean
Begin
   Q:= ∅
      For each i ∈ X do
         For each j | C_ij ∈ C do
            If Revise2000 (i, j, False) then
               If D_i = ∅ then
                  Return False
               Else
                  Q:= Q ∪ {i}
                  Return Propagation2000(Q)
               End if
            End if
         End for
      End for
End
```

Algorithm 2.18. *The AC2000 function*

```
Function propagation2000 (Q): Boolean
Begin
      While Q ≠ ∅ do
         Pop (j) from Q
         LazyMode = | Δ(j) | < ratio * | D_j |
         For each i | C_ij ∈ C do
            If Revise2000 (i, j, LazyMode) then
               If D_i = ∅ then
                  Return False
               End if
                  Q = Q ∪ {i}
            End if
         End for
         Reset Δ(j)
      End While
      Return True
End
```

Algorithm 2.19. *The Propagation2000 function*

```
Function Revise2000 (i, j, LazyMode): Boolean
Begin
    CHANGE:= False
    For each a ∈ D_i do
        If ¬ LazyMode or ∃ b ∈ Δ(j) | C_ij ∈ C then
            If ∄ b ∈ D_j | C_ij ∈ C then
                Delete a from D_i
                Add a to Δ(j)
                CHANGE:= True
            End if
        End if
    End for
    Return (CHANGE)
End
```

Algorithm 2.20. *The Revise2000 function*

The following phases illustrate the progress of the AC2000 applied to our graph-coloring example:

i	Cij	Revise 2000	*Di*	Δ(*j*)/*Q*	Propagation2000			
1	C_{12}	FALSE	D_1 = {R, G, B}	–	–			
	C_{13}	TRUE	D_1 = {R, B}	Δ(3) = {G} *Q* = {1}	C_{21}	–		
					C_{31}	–		
2	C_{21}	FALSE	D_2 = {R, G}	–	–			
	C_{23}	TRUE	D_2 = {R}	Δ(3) = {G} *Q* = {2}	C_{12}	D_1 = {B} Δ(2) = {R} *Q* = {2, 1}	C_{21}	-
							C_{31}	-
					C_{32}	–		
3	C_{31}	FALSE	D_3 = {G}	–	–			
	C_{32}	FALSE	D_3 = {G}	–	–			
			D_1 = {B}; D_2 = {R}; D_3 = {G}					

It should be noted that the storage of a support for a value a in D_j, the last time that the constraint was revised, allows us, on the one hand, to verify whether the support of a was removed or not and, on the other hand, to reduce the values to be verified during the search for a new support (there are only the values of D_j that are "after" this last support, which will be explored). [BES 01] explored this observation to improve AC2000 moving toward the AC2001 algorithm.

2.2.9. *AC2001*

In AC2001 [BES 01], the structure of $\Delta(j)$ is replaced by *Last*(i, a, j) to store the support last found for (i, a) in D_j. AC2001 assumes a certain order "$<d$" between the values of the variable domains. It is a generic algorithm, which can be used with a value or constraint-oriented scheme.

For our example, if the value $(i, 11)$ is removed, AC2001 verifies for each value a of the X_i domain that *Last* (i, a, j) still exists in D_j. It should be noted that the value $(i, 11)$ only lost its last support. AC2001 must then seek a new support for $(i, 11)$ in X_j "after" $(j, 11)$. There is no value after $(j, 11)$ so no consistency test is carried out. In this example, AC2001 requires 65 fewer tests than AC-3.

The spatial complexity of AC2001 is in $O(nd)$ while its time complexity is in $O(nd^2)$.

```
Function AC2001(X): Boolean
Begin
    Q:= ∅
    For each i ∈ X do
        For each j | C_ij ∈ C do
            If Revise2001 (i, j, False) then
                If D_i = ∅ then
                    Return (False)
                End if
                Q:= Q ∪ {i}
            End if
        End for
    End for
    Return Propagation2001(Q)
End
```

Algorithm 2.21. *The AC2001 function*

```
Function propagation2001(Q): Boolean
Begin
    While Q = ∅ do
        Pop j from Q
        For each i | C_ij ∈ C do
            If Revise2001 (i, j) then
                If D_i = ∅ then
                    Return False
                End if
                Q:= Q ∪ {i}
            End if
        End for
        Reset Δ(j)
    End while
    Return True
End
```

Algorithm 2.22. *The Propagation2001 function*

```
Function Revise2001 (i, j, LazyMode): Boolean
Begin
    CHANGE:= False
    For each a ∈ D_i do
        If ˥ LazyMode or Last (i, a, j) ∈ Δ(j) then
            If ∃ b ∈ D_j | b > d, Last (i, a, j) and C_ij ∈ C then
                Last (i, a, j): = b
            Else
                Delete a from D_i
                Add a to Δ(i)
                Change:= True
            End if
        End if
    End for
    Return CHANGE
End
```

Algorithm 2.23. *The Revise2001 function*

Below, we present the different steps of the AC2001 procedure applied to our initial example:

Phase	*i*	*Cij*	Revise2001	*Di*	Δ(*i*)/*Q*
R E	**1**	C_{12}	FALSE	D_1 = {R, G, B}	–
V I		C_{13}	TRUE	D_1 = {R, B}	**Δ(1)** = {G} **Q** = {1}
S	**2**	C_{21}	FALSE	D_2 = {R, G}	–
E		C_{23}	TRUE	D_2 = {R}	**Δ(2)** = {G} **Q** = {1, 2}

	3	C_{31}	FALSE	$D_3 = \{G\}$	–
		C_{32}	FALSE	$D_3 = \{G\}$	–
$D_1 = \{R, B\}; D_2 = \{R\}; D_3 = \{G\}; Q = \{1, 2\}$					
PROPAGATION	1	C_{21}	FALSE	$D_2 = \{R\}$	–
		C_{31}	FALSE	$D_3 = \{G\}$	–
	2	C_{12}	TRUE	$D_1 = \{B\}$	$\Delta(1) = \{G, R\}$ $Q = \{2, 1\}$
		C_{32}	FALSE	$D_3 = \{G\}$	–
	1	C_{21}	FALSE	$D_2 = \{R\}$	–
		C_{31}	FALSE	$D_3 = \{G\}$	–
$D_1 = \{B\}; D_2 = \{R\}; D_3 = \{G\}$					

As the consistency reinforcement techniques play a very important role in simplifying and accelerating resolution, numerous algorithms have been proposed in the related state of the art to enhance and/or complete these AC-i. We include Forward checking denoted by AC-1/4 [BAC 95, HAR 80], the Partial look-ahead [HAR 80], the Full look-ahead [HAR 80], the Real full look [JÉG 91, NAD 88] and MAC for "Maintaining Arc Consistency" [BES 96, SAB 94]. We also include AC-3.1 [ZHA 01], AC-3.2, AC-3.3 [LEC 03], AC-3r (AC-3 with unidirectional residues) and AC-3rm (AC-3 with multidirectional residues) [LEC 07b], which are based on the notion of residual support and which represent an improvement on the standard AC-3.

2.3. Bibliography

[ALL 94] ALLIOT J.M., SCHIEX T., *Intelligence Artificielle et Informatique Théorique*, Cépaduès-Edition, 1994.

[BAC 95] BACCHUS F., GROVE A., "On the forward-checking algorithm", *Proceedings of Principles and Practice of Constraint Programming* (*CP'95*), Lecture Notes in Artificial Intelligence, Springer-Verlag, vol. 976, pp. 292–309, 1995.

[BES 93] BESSIÈRE C., CORDIER M., "Arc-consistency and arc-consistency again", *Proceedings of National Conference on Artificial Intelligence*, Washington, D.C., USA, AAAI Press, pp. 108–113, 1993.

[BES 94] BESSIÈRE C., "Arc-consistency and Arc-consistency again", *Artificial Intelligence*, vol. 65, pp. 179–190, 1994.

[BES 95] BESSIÈRE C., FREUDER E.C., RÉGIN J.C., "Using inference to reduce arc-consistency computation", *Proceeding of IJCAI'95*, Montreal, Canada, pp. 592–598, 1995.

[BES 96] BESSIÈRE C., RÉGIN J.C., "Mac and combined heuristics: two reasons to forsake FC (and CBJ?) on hard problems", *Proceedings of International Conference on Principles and Practice of Constraint Programming*, LNCS, Springer, vol. 1118, pp. 61–75, 1996.

[BES 01] BESSIÈRE C., RÉGIN J.C., "Refining the basic constraint propagation algorithm", *Proceedings of IJCAI'01*, Seattle, WA, pp. 309–315, 2001.

[BES 05] BESSIÈRE C., RÉGIN J., YAP R., ZHANG Y., "An optimal coarse-grained arc consistency algorithm", *Artificial Intelligence*, vol. 165, no. 2, pp. 165–185, 2005.

[CHM 98] CHMEISS A., JÉGOU P., "Efficient path-consistency propagation", *International Journal on Artificial Intelligence Tools*, vol. 7, no. 2, pp. 121–142, 1998.

[HAN 88] HAN C., LEE C., "Comments on Mohr and Henderson's path consistency", *Artificial Intelligence*, vol. 36, pp.125–130, 1988.

[HAR 80] HARALICK R., ELLIOT G., "Increasing tree search efficiency for constraint satisfaction problems", *Artificial Intelligence*, vol. 14, pp. 263–313, 1980.

[JÉG 91] JÉGOU P., Contribution à l'Etude des Problèmes de Satisfaction de Contraintes: Algorithmes de Propagation et de Résolution, Propagation des Contraintes dans les Réseaux Dynamiques, PhD Thesis, LIRMM-USTL, Montpellier II, 1991.

[LEC 03] LECOUTRE C., BOUSSEMART F., HEMERY F., "Exploiting multidirectionality in coarse-grained arc consistency algorithms", *Proceedings of International Conference on Principles and Practice of Constraint Programming*, LNCS, Springer, pp. 480–494, 2003.

[LEC 07a] LECOUTRE C., CARDON S., VION J., "Path consistency by dual consistency", *Proceedings of International Conference on Principles and Practice of Constraint Programming*, LNCS, Springer, pp. 438–452, 2007.

[LEC 07b] LECOUTRE C., HEMERY F., "A study of residual supports in arc consistency", *Proceedings of International Joint Conference on Artificial Intelligence*, Hyderabad, India, pp. 125–130, 2007.

[MAC 77] MACKWORTH A., "Consistency in networks of relations", *Artificial Intelligence*, vol. 8, pp. 99–118, 1977.

[MOH 86] MOHR R., HENDERSON T.C., "Arc and path consistency revisited", *Artificial Intelligence*, vol. 28, pp. 225–233, 1986.

[MON 74] MONTANARI U., "Networks of constraints: fundamental properties and applications to picture processing", *Information Sciences*, vol. 7, pp. 95–132, 1974.

[NAD 88] NADEL B.A., "Tree search and arc-consistency in constraint satisfaction algorithms", in KANAL L., KUMAR V. (eds), *Search in Artificial Intelligence*, Springer-Verlag, pp. 287–340, 1988.

[ROS 76] ROSENFELD A., HUMMEL R., ZUCKER S., "Scene labeling by relaxation operations", *IEEE Transactions on Systems Man Cybernet*, vol. 6, pp. 420–433, 1976.

[SAB 94] SABIN D., FREUDER E.C., "Contradicting conventional wisdom in constraint satisfaction", *Proceedings of European Conference on Artificial Intelligence,* Amsterdam, Netherlands, pp. 125–129, 1994.

[TSA 93] TSANG E., *Foundations of Constraint Satisfaction*, Academic Press, 1993.

[VAN 92] VAN HENTENRYCK P., DEVILLE Y., TENG C.M., "A generic arc-consistency algorithm and its specializations", *Artificial Intelligence*, vol. 57, pp. 291–321, 1992.

[ZHA 01] ZHANG Y., YAP R.H.C., "Making AC-3 an optimal algorithm", *Proceedings of International Joint Conference on Artificial Intelligence*, Seattle, USA, pp. 316–321, 2001.

Chapter 3

CSP Solving Algorithms

Various techniques for solving CSP have been developed and classified into two categories: complete methods that guarantee completeness (quality) at the expense of efficiency (temporal complexity) and incomplete methods that sacrifice completeness for the sake of efficiency. A method is called complete if it manages to find a solution, if one exists, or, vice versa, if it detects the absence of a solution. In other instances, it is considered an incomplete method.

Thus, the complete methods explode in the presence of large-scale applications while the incomplete methods can find a "good quality" solution in a "reasonable" amount of time. This chapter only discusses the complete methods insofar as we are interested, here, in obtaining a solution that satisfies all the constraints.

3.1. Complete resolution methods

The algorithms discussed in this section are all based on tree search. These are improvements or extensions of the basic and most primitive algorithm "generate and test", which consists of generating all the solutions and then testing their consistency, which is done by visiting the entire search space. It should be noted that in the illustrative diagrams in the following figures, the nodes annotated with crosses (X) represent failures.

3.1.1. *The backtracking algorithm*

Faced with the combinatorial explosion of "generate and test", the idea of combining generation and test has led to the backtrack (BT) algorithm, which relies on the depth-first search and is based on the recursive BT function that takes a sequence V of variables of X to be instantiated (initially X including all the variables) and an initially empty instantiation I as arguments (see algorithm 3.1).

```
BT (V, I)
Begin
    If V = Ø then
        I is a solution
    Else
        Let x ∈ V
        For each v ∈ D_x do
            If I ∪ {(x, v)} is consistent then
                BT (V- {x}, I ∈ {(x, v)})
            End if
        End for
    End if
End
```

Algorithm 3.1. *The recursive BT procedure*

The BT algorithm's principle consists of starting from an empty instantiation and performing a tree traversal on the set of variables [FRE 82]. The instantiation of a variable x by a value v from its domain triggers a consistency verification procedure for all the constraints that involve the variable x and the variables that have already been instantiated. A failure result leads, on the one hand, to the rejection of the inconsistent partial assignment (as any assignment containing an inconsistent partial assignment is itself inconsistent) and, on the other hand, to the generation of other nodes by selecting another value v for the current variable x. This improves the performance in terms of the number of nodes visited and the number of constraint checks. For the four-queens example of the problem shown in Figure 3.1 and exploiting the property of symmetry (a queen i can only be placed on one of the four squares in column i), the number of nodes visited by *generate and test* is 256 (4^4) while for BT it is 26, in other words around six times less. However, its temporal complexity is still exponential but still far better than that of its predecessor.

The criticism most frequently directed at this algorithm mainly covers the following points: the useless exploration of unpromising subspaces, rediscovering the same local inconsistencies and failing to take into account the inconsistency of the current instantiation with variables not yet instantiated.

To overcome these shortcomings, a vast number of algorithms incorporating an "intelligent" look-back mechanism and an *a posteriori* verification of the look-ahead consistency were proposed. In the following sections, we present the best known of these. Others can be found in [GLI 04a] and [GLI 04b].

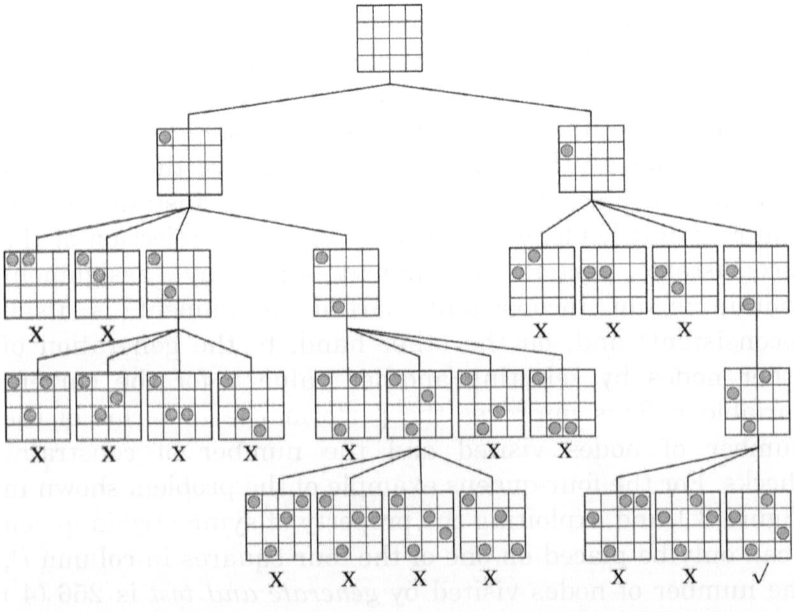

Figure 3.1. *The search tree explored by the backtrack algorithm for the four-queens problem (example 1.1, subsection 1.2.2) where* X *and* √, *respectively, represent a failure and a solution*

3.1.2. *Look-back algorithms*

Before discussing these algorithms, let us recall the common exploration scheme that is identical to that of the BT algorithm. These methods extend an initially empty assignment of the problem's variables. This assignment is extended to a new variable/value couple while respecting the constraints shared with the variables previously included in the assignment. We call these *passed* variables. The variable being examined is described as *current*. Finally, all *future* variables consist of variables not yet considered.

3.1.2.1. *The backjumping algorithm*

```
BJ (V,I)

Begin
    If V = Ø then
        I is a solution
    Else
        Let x ∈ V
        For each V ∈ D_x do
            if Consistent (I, x, V) then
                jump:= BJ (V – {x}, I ∪ {(x, v)})
                If jump ≠ i then
                    Return (jump)
                End if
            End if
        End for
        jump:= jump-place[i]
        j:= jump+I
        While j ≤ i do
            jump-place[j]:= 0
            j:= j+1
        End while
        Return (jump)
    End if
End
```

Algorithm 3.2. *The recursive BJ function*

The backjumping (BJ) algorithm constitutes one of the improvements proposed for backtracking. However, if the graph of constraints is complete, it coincides with BT.

According to [VAN 94], the BJ algorithm developed by Gashing in 1977 proceeds with direct intelligent backtracking toward the source or sources of conflict so as to avoid searching in unpromising spaces. This mechanism can reduce the number of nodes visited as well as the number of constraint checks. The process for determining sources of conflict is simple. The failure to instantiate a variable X_i leads to backtracking toward the deepest variable X_j in the tree search such that the constraint C_{ij} is not satisfied, i.e. the variable most recently instantiated that is connected to the current variable. The corresponding algorithm is based on the recursive BJ function. It calls upon the *consistent* function whose pseudocode is presented in algorithm 3.3.

```
Consistent (I, Xi, X)
Begin
Consistent:= True
For each instantiated Xj ∈ X – {Xi} do
    While consistent do
        If {(Xi, vi'), (Xj, vj')} violates the constraint Cij then
          If j > jump-place [i] then
            jump-place [i]:= j
            consistent:= False
            Return (consistent)
          End if
        End if
    End while
End for
jump-place [i]:= i-1
Return (consistent)
End
```

Algorithm 3.3. *The consistent function*

The *consistent* function that receives the current partial instantiation, the current variable and the problem set of variables defines a jump-place parameter initialized to zero for all variables, all before the search. During the consistency verification procedure (Algorithm 3.3.) and due to the failure (violation) of a constraint, the current jump-place receives the index of the variable (source of the failure) if that one is deeper in the search tree than the current value of the jump-place (lines 4–6). The BJ algorithm is applied to a sequence *V* of variables and to an initially empty instantiation *I*, and returns the index of the variable toward which a backtrack will be carried out if the instantiation is inconsistent.

3.1.2.2. *The conflict-directed backjumping algorithm*

```
Consistent -CDB (I, Xi, X)
Begin
  consistent:= True
  For each instantiated Xj ∈ X –{Xi} Do
    While consistent Do
      If {(Xi, vi'), (Xj, vj')} violates the constraint Cij then
        Conflicts[i], [j]:= 1/* There is a conflict between Xi et Xj */
        If Conflicts [i] [i] < j then
          Conflicts [i] [i]:= j
          consistent:= False
          Return (consistent)
        End if
      End if
    End while
  End for
  Return (consistent)
End
```

Algorithm 3.4. *The consistent-CDB function*

This algorithm improves on the BJ algorithm, which aims to find an appropriate variable to BT [PRO 93]. Indeed, following a backjump by the variable X_i toward X_j, conflict-directed backjumping (CDB) continues to carry out backjumps through the conflicts that involve both X_i and X_j. This mechanism is ensured by saving the conflict set relating to each variable. During resolution, this initially empty set receives all of the instantiated variables inconsistent with the current variable. The failure to instantiate a variable X_i causes backtracking toward the most recently instantiated variable X_k in its conflict set. This changes the conflict set of X_j, which consequently unites the current conflict set of the variable X_j and of that for the variable X_i. The algorithm is embodied in the recursive CBJ function that takes a sequence V of variables to be instantiated and an initially empty instantiation I as an argument, and which returns a set of variables called *conflict set*. It calls upon the consistent-CDB function whose pseudocode is presented in algorithm 3.5.

```
Union-Conflict(k, j)

Begin

    m:= I
    While m < k do
      Conflicts [k] [m]:= Conflicts [k] [m] or Conflicts [j] [m]
      If (Conflicts [k] [m] & Conflicts [k] [k] < m) then
        Conflicts [k] [k]:= m
      End if
      m:= m + 1
    End while

End
```

Algorithm 3.5. *The union-conflict procedure*

The *conflict* [*i*][*i*] structure receives the index of the deepest variable in the search tree that belongs to the conflict set of the variable X_i toward which the jump will be performed.

The Conflict based Jumping (CBJ) function also calls upon the *union-conflict (jump, i)* procedure that determines the union of the conflict sets for the variables X_{jump} and X_i.

CBJ (V, I)

Begin

```
If V = ∅ then I is a solution
  count:= count + 1/* count corresponds to the solutions number*/
Else
  curr:= count
  Let x ∈ V
  For each V ∈ D_x do
    If Consistent-CBD (I, x, V) then
      jump:= CBJ (V-{x}, I ∪ (x, v))
      If (jump ≠ i) then
        Return (jump)
      End if
    End if
  End for
  If (curr = count) then
    jump:= Conflicts [i] [i]
  Else
    jump:= i-I
  End if
  Union-Conflict(jump, i)
  Empty-Conflict(jump, i)/* Delete those sets for all
                   variables between jump and I */
  Return (jump)
End if
```

End

Algorithm 3.6. *The recursive CBJ function*

The variable *curr* is defined to test whether a solution has been determined in the current search space. If so, the algorithm examines other instantiations for the current variable that are followed by a BT toward the parent node. If no solution has been detected, the algorithm proceeds by backtracking toward the source of the failure.

This approach in some sense corresponds to backtracking controlled by dependencies. According to [JÉG 91], such approaches were also proposed by Stallman and Sussman [STA 77] and Doyle [DOY 79].

3.1.2.3. *The graph-based-backjumping algorithm*

This algorithm was introduced by R. Dechter in 1990 [DEC 90]. With each failure, a BT is performed toward the connected variable of the constraint graph that is the closest in order of instantiation. The method does not isolate the variables in conflict with the current variable.

During a BT phase starting from the current variable i, the algorithm "jumps" toward the connected variable h that is closest in order of instantiation. From there, the next jump is directed toward the closest passed variable connected with i and/or h. If there is no connected passed variable, the problem's inconsistency is detected.

3.1.2.4. *The dynamic backtracking algorithm*

This algorithm was introduced in [GIN 93]. It starts from the observation that the BJ algorithm ignores all of the intermediary work carried out when BJ occurs from a failing variable i toward a variable j that has already been instantiated. The intermediary assignments k, such as $j < k \leq i$, are lost. The second observation is that if the current domain of k is reduced due to the assignment of the previous variables in the order of instantiation, this reduction can be conserved by taking the future change in the instantiation of j into account.

Thus, dynamic backtracking can be seen as an improvement of the previous algorithms. It records the same information as CDB, but in more detail. At each step and for each variable x and each value v that is eliminated from the domain of x, it records the set of previously assigned variables that are responsible for this elimination. These sets are called: *eliminating explanations* and the space necessary to record them is around $O(n^2d)$, where n is the number of variables and d is the maximum size of domains. In practice, during the instantiation of a variable x, the domain D_x of this variable is reduced by removing values that are incompatible with the assignments earlier in the order of instantiation. Thus eliminating explanation E_x is constructed. It contains the reasons for removing x values from the current domain. Should the instantiation of x fail, E_x can return toward the closest variable h involved in the conflict. Some method jumps are backward rather than forward, but they conserve the intermediary eliminating explanations. However, these removals cause the method to successively recalculate the same explanations of failures.

3.1.2.5. *The informed-backtrack algorithm*

The informed-backtrack algorithm, referred to as IBt, combines conflict-minimizing heuristics (better known as min-conflicts heuristics) with the simple BT principle [MIN 92]. This algorithm constitutes a repair method, which has been used successfully on a variety of benchmarks [MIN 92, TSA 96].

```
Informed-Backtrack (V, D, C)
Begin
    I:= Ø
    R:= Ø
    For each x ∈ V do
      Let v ∈ D_x
      I:= I ∪ {(x, v)}
    End for
    IBT (I, R, D, C)
End
```

Algorithm 3.7. *The informed-backtrack algorithm*

The IBt calls upon the recursive IBt procedure after having constructed the set of random instantiations *I*.

```
Procedure IBT (I, R, D, C)

Begin
    If Conflicts (I) then
      x:= take-var (I)
      list:= Ordervalues (x, D_x, I, R, C)
      While (not empty (list)) Do
        w:= first (list)
        R:= R ∪ {(x, v)}
        Result = IBT (I- {(x, v)}, R, D, C)
        If Result ≠ Nul then
          Return (Result)
        End if
      End while
      Return Nul
    Else
      Return I ∪ R
    End if
End
```

Algorithm 3.8. *The recursive IBt procedure*

The IBt procedure considers two sets I and R. I is initialized to a set of random instantiations for all the variables and R is initialized to the empty set in which the IBt places the consistent partial instantiation. The procedure then starts to solve the conflicts that exist. If there is a conflict between the couple (x, v) and another instantiation in I, then this couple will be dealt with and then removed from the set I.

For all the values v' such that (x, v') is compatible with all the instantiations in R, v' will be placed in a list in increasing order of the number of conflicts that it causes with the instantiations of I. Then, the value with the lowest number of conflicts will be assigned to x and this instantiation will be added to R. If such a value does not exist (all the instantiations of x have conflicts with some instantiations in R), then there will be a BT. The process terminates in one of the two cases: (1) no conflict is detected or (2) all combinations of instantiations have been tried. IBt ensures completeness and does so by visiting all the branches of the search space, if necessary.

The *values-order* procedure puts the values of a given variable x into order according to the increasing order of the number of conflicts that this variable can have. It should be noted that this procedure only considers the values that can take x while keeping the consistent instantiation. This procedure always favors the best values: those that violate the minimum of constraints with the future variables and that are consistent with the already instantiated variables.

```
Procedure Order-values (x, D_x, I, R, C)

Begin
    List:= Ø
    For each v ∈ D_x Do
        If Consistent ({(x, v)}, R) then
            nbcv:= 0
        End if
        For each {(y, w)} ∈ I Do
            If (Not consistent ({(x, v), (y, w)}, C_xy) then
                nbcv:= nbcv+1
            End if
        End for
        List:= List ∪ {(v, nbcv)}
    End for
    Queue:= sort-in-increasing order (List)
    Return (Queue)
End
```

Algorithm 3.9. *The values-order procedure*

3.1.3. *Look-ahead algorithms*

Filtering is also applied while seeking solutions. It is used during the construction of a partial assignment of variables to reduce the remaining search space, and does so on the basis of the commitment undertaken (assignments already performed). This, therefore, corresponds to forward filtering, carried out in the search space relating to the future variables.

The following methods combine a solutions search and filtering at various levels. We limit our discussion to arc consistency and assume all inter-variable constraints to be binary.

3.1.3.1. *Forward checking*

```
FC (V, I)
Begin
    If V = Ø then
        I is a solution
    Else
        Let x ∈ V
        For each v ∈ D_x Do
            Push (V – {x})
            If Check-Forward (x, v, V) then
              FC (V – {x}, I ∪ {(x, v)})
            End if
            Pop (V– {x})
        End for
    End if
End
```

Algorithm 3.10. *The recursive FC procedure*

In the domain D_i of the not yet instantiated future variables, the filtering process consists of removing the values that are inconsistent with the current instantiation and therefore corresponds to a propagation that limits itself to the neighborhood of the current variable and does so due to complexity [BAC 95]. Thus, this algorithm can, if there are any, detect inconsistencies between the current variable and the future variables.

The forward-checking (FC) algorithm is embodied in the recursive FC procedure that takes a sequence V of variables to be instantiated and an initially empty instantiation I as arguments. It calls upon the check-forward function whose pseudocode is presented in algorithm 3.11.

```
Check-Forward (X_i, v, V)
Begin
      consistent:= True
      For each X_j ∈ V – {X_i} do
        While consistent do
          For each v' ∈ D_j do
            If {(X_i, v), (X_j, v')} violates the constraint C_ij then
              D_j:= D_j – {v'}
            End if
          End for
          If D_j = Ø then
            consistent:= False
          End if
        End while
      End for
    Return (consistent)
End
```

Algorithm 3.11. *The check-forward function*

To illustrate this algorithm, we present, in Figure 3.2, the search tree traversed for the resolution of the four-queens problem described in subsection 1.2.2. This figure shows the benefits of this algorithm compared with BT, in terms of the number of constraint checks and the number of visited nodes that goes from 26 to 7, which shows an important gain in terms of temporal complexity even if this complexity remains, in the worst case, exponential.

For several years, FC was irreproachable with regard to its efficiency in solving CSPs. Numerous variants of this algorithm have been developed and evaluated including

FC-CBJ [PRO 93], minimal FC [DEN 94, DEN 96], minimal FC with backmarking [KWA 96a] and minimal FC with backmarking and CDB [KWA 96b]. However, this reputation has been challenged by a new algorithm denoted by maintaining arc consistency (MAC) whose performance far exceeds that of FC and FC-CBJ [BES 96].

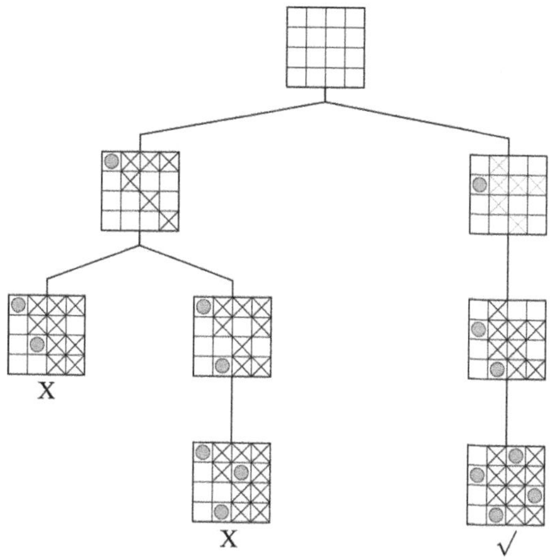

Figure 3.2. *Resolution of the four-queens problem via forward checking*

3.1.3.2. *The arc-consistency look-ahead algorithm*

Presented several times in particular by [GAS 79], as "domain element elimination with backtracking" (DEEB), this algorithm verifies during each instantiation that the set of future variables satisfies the arc-consistency property. During the instantiation of the current variable i with the value v, the future variables j have their domains filtered according to what has just been carried out. For this, the algorithm calls upon one of the arc-consistency procedures with the set of future variables as an argument.

3.1.3.3. *Maintaining arc consistency*

The MAC algorithm has appeared under several names. It was initially proposed under the name CS2 in [GAS 74], then reformulated and extended to DEEB in 1979 [GAS 79]. It has the same basic principles as FC, alternating consistency maintenance and solution generation [SAB 94]. However, instead of maintaining a low level of consistency resulting from the establishment of the *Revise* procedure, from the variables not yet instantiated to the variable that has just been instantiated (identical to that of FC), MAC maintains the full arc consistency established by the removal of all inconsistent values and the propagation of the side effects related to this elimination. Note that this propagation is not limited to the neighborhood of the current variable, as in the FC case. Instead, it takes all interactions between its neighbors into account.

The principle of this algorithm is simple. It occurs in two phases [PRO 95, SAB 94]:

– *Phase 1*: MAC makes considerable effort to obtain an initially arc-consistent graph of constraints and does so by executing an AC-*i* procedure.

– *Phase 2*: This phase is the resolution phase. During the progress of the algorithm, the instantiation of a variable X_i by a value val_i leads to the removal of all remaining values in D_i. The effect of this elimination is propagated through the graph of constraints to restore a full arc consistency. The failure to instantiate a variable leads to a return towards the parent node and the restoration of future variable domains.

This means that MAC carries out more work in terms of constraint checks since it carries out more arc revisions. With this additional work, the method develops fewer nodes, which can clearly be seen in Figure 3.3 showing the search tree traversed by MAC to solve the simple four-queens problem.

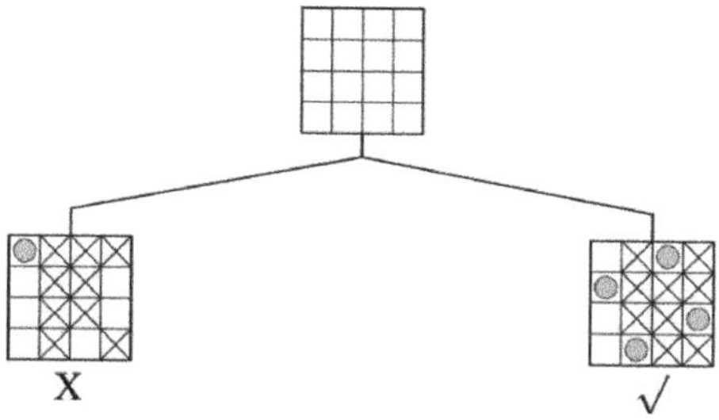

Figure 3.3. *Resolution of the four-queens problem via MAC*

```
AC3-MAC (F, X_cour)
Begin
    consistent:= True
    While ((F ≠ Ø) & (consistent)) do
       Select and Delete from F a pair (X_i, X_j)
       If Revise (X_i, X_j) = True then
          consistent:= not empty (D_i)
          F:= F ∪ {X_k, X_i} | ∃ a constraint C_ki & k ≠ i&k > cour
       End if
    End while
    Return (consistent)
End
```

Algorithm 3.12. *AC-3-MAC*

There have been several versions of AC-3-MAC, which rely on an AC-3 procedure to establish full arc consistency [PRO 95]. Other versions that call upon other AC-*i* procedures have been proposed in the related literature [SAB 94, SAB 97].

The rediscovery of MAC by Sabin and Freuder [SAB 94] has pushed research in this new direction which has turned out to be interesting in terms of complexity. Consequently, refinements improving the performance of the method [SAB 97] as well as hybridizations mixing it with an intelligent mechanism [PRO 95] are proposed. Readers interested in a more comprehensive survey should refer to the bibliography cited.

3.1.3.4. *The backmarking algorithm*

This strategy proposed by [GAS 77] belongs to those that proceed with an *a posteriori* verification of the consistency. It aims to eliminate the repetition of consistency tests that have already been carried out. During the instantiation of X_k, if backtracking is necessary up to X_i the procedure records the values of X_k that are consistent with those assigned to the variables from X_1 to X_{i-1}. Thus, after a new instantiation of the variables from X_i to X_{k-1}, in X_k, the only consistency tests necessary will be carried out in relation to variables from X_i to X_{k-1}. A more detailed description is presented in [VAN 94].

3.2. Experimental validation

CSPs are combinatorial problems known for being NP hard. The resolution methods, previously discussed, explode with the size of the problem dealt with. Since they are all exponential in the worst case, several comparison criteria have been proposed, mainly to estimate the search cost, which is often made on the basis of randomly generated problems. From these criteria, we distinguish resolution time or CPU time, the number of constraint checks and the number of nodes visited.

3.2.1. *Random generation of problems*

Although early work on CSP compared how the methods solved particular problems (such as the n-queens problem or the graph-coloring problem), very soon people were interested in their experiments with randomly generated problems where the problem's parameters can be controlled and varied. Often, the tests focused on binary CSP problems, whose relations are expressed in extension.

Generation of the problems is achieved on the basis of the quadruplet (n, d, p_1, p_2) where:

– n is the number of variables for the problem;

– d is the number of values (size) for each domain;

– p_1 is the probability that a constraint links a pair of variables;

– p_2 is the conditional probability that a pair of values is inconsistent if there is a constraint between the associated variables.

Here p_1 and p_2 represent the density and tightness of the constraints, respectively. The density and tightness are between 0 and 1. For example, for a given binary constraint C_i, tightness is represented by the ratio for *the number of couples forbidden by C/total number of couples*. Thus the closer this tightness gets to 1, the more difficult it is to satisfy this constraint and conversely, the closer it is to 0, the easier it is to satisfy. There are several ways to determine these probabilities. Overall, there are two main models [TSA 98]:

– *Model A*: We select each of the possible $(n(n-1)/2)$ arcs from the graph of constraints, independently, with a p_1 probability. Then, for each couple of variables linked by a constraint, we go through the d^2 couples of their values and mark each of them inconsistent with a p_2 probability. With this method, there will be a set of problems for which we have, on an average, $p_1(n(n-1)/2)$ constraints and p_2d^2 inconsistent couples for each constraint. However, for a set of randomly generated problems, there can be considerable variations. This may alter the results because the generated problems may be heterogeneous even if they have the same parameters. That is why we should rather generate a large number of problems.

– *Model B*: It consists of generating problems for which p_1 and p_2 specify, respectively, the number of constraints and the number of inconsistent value pairs. We then speak of proportions: the method chooses from the possible constraints, precisely $p_1(n(n-1)/2)$ (rounded to the nearest integer) randomly drawn constraints; and for each constraint, it generates precisely p_2d^2 couples of inconsistent values. This model is increasingly used in the experiments, due to the homogeneity of the problems that it generates.

Although, in general, all the domains have the same size and all the constraints have the same tightness, the generator can be easily changed by defining an interval $I_{\text{size}} = [d - \Delta_t, d + \Delta_t]$ for the domains and an interval $I_{\text{tightness}} = [p_2 - \Delta_d, p_2 + \Delta_d]$ for the constraints. For each domain, its size is randomly chosen from I_{size} with a uniform probability. And, in the same way, the tightness of each constraint is chosen from $I_{\text{tightness}}$.

Furthermore, it should be noted that it would be preferable, in the experiments, if the graph of constraints is connected (cannot be broken down in disconnected sub-graphs). The CSPs with graphs of constraints not connected are solved with branch and bound strategies.

3.2.2. *Phase transition*

Phase transition is a phenomenon that was observed in a lot of NP-hard problems. The transition is made from the region where the problems are sub-constraint, and therefore relatively easy to solve, toward the region of the over-constrained problems for which it is relatively easy to prove insolubility. This phenomenon was clearly observed for the random binary CSP problems. Indeed, by setting (n, d, p_1), varying the tightness p_2 and applying a resolution method, we noted three regions (see Figure 3.4).

Region I corresponds to the sub-constraint problems, i.e. whose tightness is weak, and are therefore easy to solve. Region III corresponds to the over-constrained problems, i.e. whose tightness is high, and are therefore difficult to solve. In region II, the problems are much more difficult to solve. This region corresponds to the phase transition. This is called the “mushy region”. [TSA 98] shows that the peak is when 50% of the problems are solvable and when the average number of solutions for these problems is equal to 1. Estimation was proposed for the tightness p_{2crit} for this critical type of problem: ${}_{2crit} = 1 - d^{-2/p1(n-1)}$

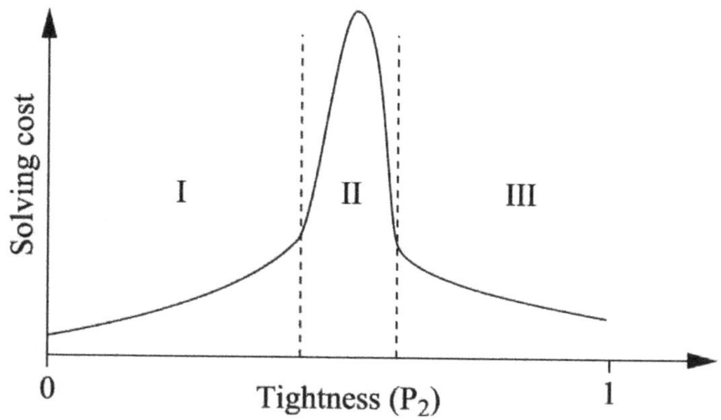

Figure 3.4. *Phase transition*

Note that this phase transition phenomenon is not observed for the methods that seek all of the solutions to the problem, since they must make a tiresome traversal through the first region of the easy problems to find all the solutions.

The mushy region remains very important in light of its difficulty and the fact that many real problems are in this region. Much work is based on the results obtained in this region for the sake of comparison.

3.3. Bibliography

[BAC 95] BACCHUS F., GROVE A., "On the forward-checking algorithm", *Proceeding of Principles and Practice of Constraint Programming (CP'95)*, Lecture Notes in Artificial Intelligence, vol. 976, Springer-Verlag, pp. 292–309, 1995.

[BES 96] BESSIÈRE C., RÉGIN J.C., "Mac and combined heuristics: two reasons to forsake FC (and CBJ) on hard problems", *2nd International Conference on Principles and Practice of Constraint Programming*, LNCS, Springer, pp. 61–75, 1996.

[DEC 90] DECHTER R., "Enhancement schemes for constraint processing: backjumping, learning and cutset decomposition", *Artificial Intelligence*, vol. 41, pp. 273–312, 1990.

[DEN 94] DENT M., MERCER R., "Minimal forward checking", *Proceedings of the International Conference on Tools with Artificial Intelligence*, IEEE Computer Society, pp. 432–438, 1994.

[DEN 96] DENT M., MERCER R., "An empirical investigation of the forwardchecking algorithm and its derivatives", *Proceedings of International Conference on Tools with Artificial Intelligence*, Toulouse, France, 1996.

[DOY 79] DOYLE J., "A truth maintenance system", *Artificial Intelligence journal*, pp. 231-272, 1979.

[FRE 82] FREUDER E.C., "A sufficient condition for backtrack-free search", *Journal of the ACM*, vol. 29, pp. 24–32, 1982.

[GAS 74] GASHING J., "A constraint satisfaction method for inference making", *Proceedings of the 12th Annual Allerton Conference on Circuit and System Theory*, University of Illinois, Urbana, IL pp. 866–874, 1974.

[GAS 77] GASHING J., "A general backtracking algorithm that eliminates most redundant tests", *Proceedings of the International Joint Conference on Artificial Intelligence*. Cambridge, MA, USA, 1977.

[GAS 79] GASHING J., Performance measurement and analysis of certain search algorithms, PhD Thesis, Carnegie-Mellon University, 1979.

[GIN 93] GINSBERG M.L., "Dynamic backtracking", *Journal of Artificial Intelligence Research*, vol. 1, pp. 25–46, 1993.

[JÉG 91] JÉGOU P., Contribution à l'Etude des Problèmes de Satisfaction de Contraintes: algorithmes de Propagation et de Résolution, Propagation des Contraintes dans les Réseaux Dynamiques, PhD Thesis LIRMM-USTL, Montpellier II, France, 1991.

[JLI 04a] JLIFI B., GHÉDIRA K., "On the enhancement of the informed backtracking algorithm", *Proceedings of International Conference on Principles and Practice of Constraint Programming*, LNCS, Springer, 2004.

[JLI 04b] JLIFI B., GHÉDIRA K., "On the enhancement of the informed backtracking algorithm, principles and practice of constraint programming", *Proceedings of the 2nd European Starting AI Researcher Symposium in conjunction with European Conference on Artificial Intelligence*, E. Omaindia and S. Staab Edts, IOS Press, Amsterdam, The Netherlands, 2004.

[KWA 96a] KWAN A., TSANG E.P.K., Minimal forward-checking with backmarking, Technical Report CSM-260, University of Essex, Colchester, UK, March, 1996.

[KWA 96b] KWAN A., TSANG E.P.K., "Minimal forward-checking with backmarking and conflict-directed-backjumping", *Proceedings of the Conference on Tools with Artificial Intelligence,* Toulouse, France, 1996.

[MIN 92] MINTON S., JOHNSON M.D., PHILIPS A.B., LAIRD P., "Minimizing conflicts: a heuristic method for constraint-satisfaction and scheduling problems", *Artificial Intelligence*, vol. 58, pp. 161–205, 1992. [Reprinted in "Constraint-based Reasoning", FREUDER E.C., MACKWORTH, A.K. (eds), MIT Press, 1994.]

[PRO 93] PROSSER P., "Hybrid algorithms for the constraint satisfaction problems", *Computational Intelligence*, vol. 9, no. 3, pp. 268–299, 1993.

[PRO 95] PROSSER P., Mac-CBJ: maintaining arc-consistency with conflict-directed backjumping, Research Report/95/177, Department of Computer Science, University of Strathclyde, Glasgow, Scotland, 1995.

[SAB 94] SABIN D., FREUDER E.C., "Contradicting conventional wisdom in constraint satisfaction", *Proceedings of European Conference on Artificial Intelligence*, Amsterdam, The Netherlands, pp. 125–129, 1994.

[SAB 97] SABIN D., FREUDER E.C., "Understanding and improving the MAC algorithm", *Proceedings of the International Conference on Principles and Practice of Constraint Programming*, LNCS, Springer, 1997.

[STA 77] STALLMAN R.M; SUSSMAN G.J., "Forward reasoning and dependency-directed backtracking in a system for computer-aided circuit analysis". *Artificial Intelligence Journal*, pp. 135–196, 1977.

[TSA 96] TSANG E.P.K., *Foundations of Constraint Satisfaction Problems*, Department of Computer Science, University of Essex, Colchester, Essex, UK, 1996.

[TSA 98] TSANG E.P.K., WANG C.J., DAVENPORT A., VOUDOURIS C., LAU T.L., *A Family of Stochastic Methods for Constraint Satisfaction and Optimization*, University of Essex, Colchester, UK, 1998.

[VAN 94] VAN RUN P., Domain independent heuristics in hybrid algorithms for CSP, PhD Thesis, University of Waterloo, Canada, 1994.

Chapter 4

Search Heuristics

Given the NP-complete aspect of CSPs and their extensions, despite the numerous techniques proposed within this framework, other directions have been explored for speeding up the resolution. The first direction is based on the search-space organization (parallel or distributed search) and the second on the instantiation order according to variables or values or even constraints.

4.1. Organization of the search space

4.1.1. *Parallel approaches*

The parallel approach is different from the distributed approach, which will be discussed in Chapter 8. It aims to accelerate the resolution via simultaneous handling of different processing units. However, the distributed approach is confronted with a situation where the problem is spread over these units in many different subproblems [HAM 99].

In the main approaches for seeking solutions via parallel backtracking, the space explored by the different processing units is disjoint [RAO 93]. Each execution node is in charge of exhaustively exploring its subspace. In this case, it does not need to coordinate with the other units. This implies that each node possesses the entire problem. A duplication phase is then necessary before initiating the search.

In general, the explored subspaces may not be uniform even if they are of the same size. Some subspaces will be explored more quickly than others. Thus, a mechanism for dynamic partitioning is frequently introduced to cope with load imbalances caused by disparity between the subspaces.

With this mechanism, a processor that has finished its work sends a work request to a processing unit that is still occupied. One of the strategies proposed for the required unit is to divide the remaining search space into two subspaces and to send one of them to the requester [RAO 93]. As each processor possesses the entire problem it does not receive subproblems, but regions of the search space.

4.1.2. *Distributed approaches*

From the numerous distributed approaches, we choose to discuss the following:

The constraint partition and coordinated reaction (CP&CR) approach, presented by J. Liu and K.P. Sycara [LIU 95], is specific to the distributed resolution of scheduling problems. Entities called "agents" are each assigned a subproblem. Each subproblem consists of a set of constraints of the same type. Each agent is in charge of instantiating the variables of its subproblem. Furthermore, the shared variables can be modified by the set of agents that involve them. Coordination is carried out using various strategies. Note that the system is not guaranteed to find a solution even if a solution exists. Finally, if the problem is unsolvable, stopping of the interactions must be triggered by an external entity.

The multiagents for the satisfaction of constraints (MASC) approach, presented by K. Ghedira [GHE 93], is principally based on a multi-agent adaptation of an incomplete simulated-annealing resolution method. The obtained model – constraint agents and variable agents – differs completely from the distributed resolution approaches of CSPs. Constraint agents and variable agents, in direct interactions, seek to achieve their maximum satisfaction. A constraint is satisfied if it is instantiated and if its relation is verified. A variable is satisfied if it is instantiated. It is all the more satisfied if it satisfies constraints. This approach makes full use of the multi-agent nature and does not use any existing CSP resolution technique. Indeed, in terms of the resolution method, there is neither a tree search nor a backtrack but rather a movement within the space of configurations to reach an optimum configuration that corresponds to a solution if the CSP is solvable. This movement is based on the simulated-annealing technique that consists of:

– allowing for local satisfaction losses, for each variable, in the hope that they bring an overall gain;

– managing this possibility using the temperature parameter whose value is high at the beginning thus allowing for a loss in satisfaction. This parameter then goes through a very slow and progressive decrease until the system is stabilized.

The DisCSP approach: In [YOK 92], the authors define the DisCSPs as distributed problems. It is therefore no longer a matter of standard CSPs that are distributed in the hope for more effective processing. Their approach belongs to the cooperative distributed problem solving (CDPS) framework. This particular area of AI seeks to solve problems by coordinating artificial agents. The authors informally define DisCSPs as binary CSPs where the variables are distributed between different agents. These agents communicate by sending messages and possess all information relating to their variables: the domains and the constraints involving these variables. These DisCSPs will be broadly discussed in Chapter 8.

4.1.3. *Collaborative approaches*

Collaborative research constitutes a parallelization strategy in one sense, and a distribution strategy in another [HAM 99]. Indeed, it implements various agents in charge of solving all or part of the same problem. These agents possess all or part of the data relating to the problem and implement various search strategies.

Collaboration begins when various agents exchange and reuse the information that is useful for the search process. These exchanges correspond to suggestions between agents. It is at this level that these approaches are similar to distribution as we have shown.

The main difficulty of these methods results from this collaboration. The collaborative framework must define the various search methods used by the group as well as the way in which they interact with each other. This essentially amounts to specifying which information to transmit, which mode of transmission to use and when/how to use the information received.

4.2. Ordering heuristics

4.2.1. *Illustrative example*

Given a simple CSP consisting of three variables X, Y and Z whose respective domains are: $D_x = \{5, 2\}$, $D_y = \{2, 4\}$ and $D_z = \{5, 2\}$. The problem consists of seeking a solution in such a way that the value of Z divides both those of X and Y.

This restriction can be expressed in terms of two binary constraints: the first between X and Z and the second between Y and Z. These relations are expressed in extension by:

$$R_{xz} = \{(5, 5), (2, 2)\} \text{ and } R_{yz} = \{(2, 2), (4, 2)\}.$$

The graph of constraints associated with this CSP is presented in Figure 4.1.

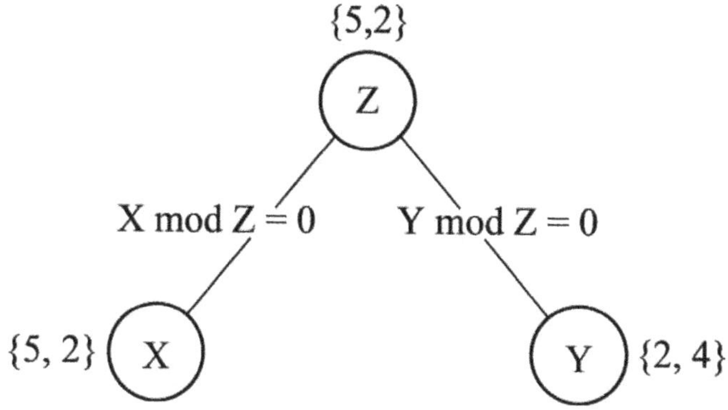

Figure 4.1. *Graph of constraints associated with the illustrative example*

The search space, explored by one of the backtracking resolution algorithms, depends on the order followed while going through the variables. For example, if we want to seek all the possible solutions and if the algorithm follows the alphanumeric order X, Y and Z, it traverses the search tree in Figure 4.2(a). On the other hand, if it follows the reverse order (Z, Y and X), then it traverses the search tree that is smaller (Figure 4.2(b)).

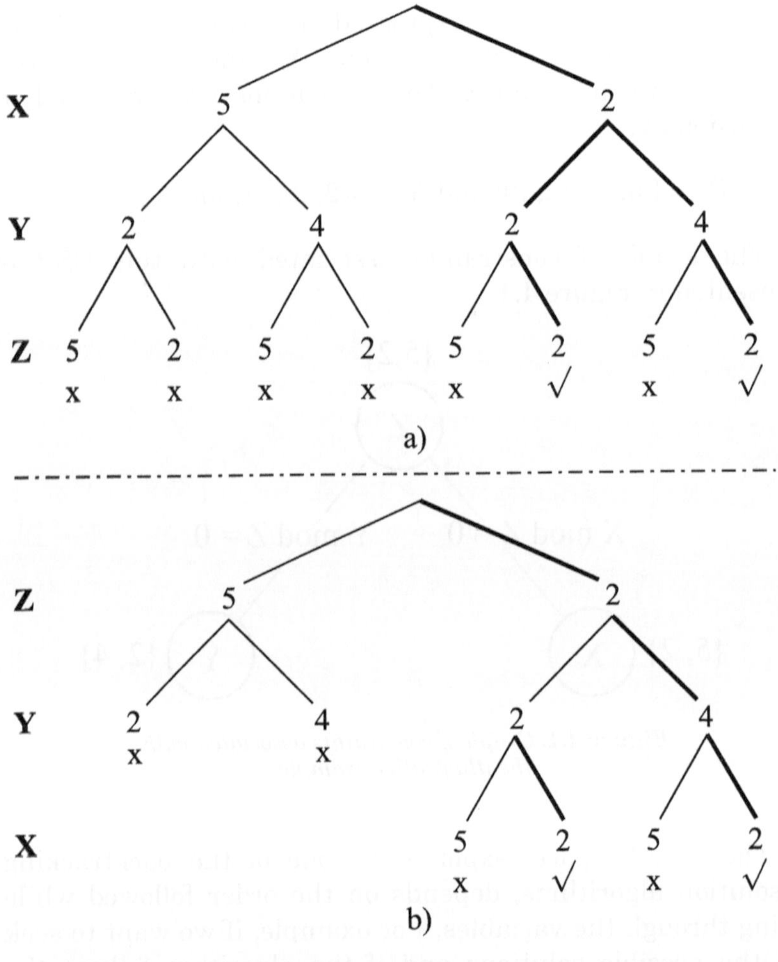

Figure 4.2. *Search trees explored by Bt for seeking all of the solutions for the illustrative example: a) according to the order X, Y* and *Z; b) according to the order Z, Y,* and *X*

When we wish to find a single solution and if the algorithm follows the alphanumeric order X, Y and Z, it traverses the search tree in Figure 4.3(a). On the other hand, if it follows the reverse order (*Z*, *Y* and *X*), it traverses the tree in Figure 4.3(b). Note that the traversal in Figure 4.3(b) is better than that in Figure 4.3(a).

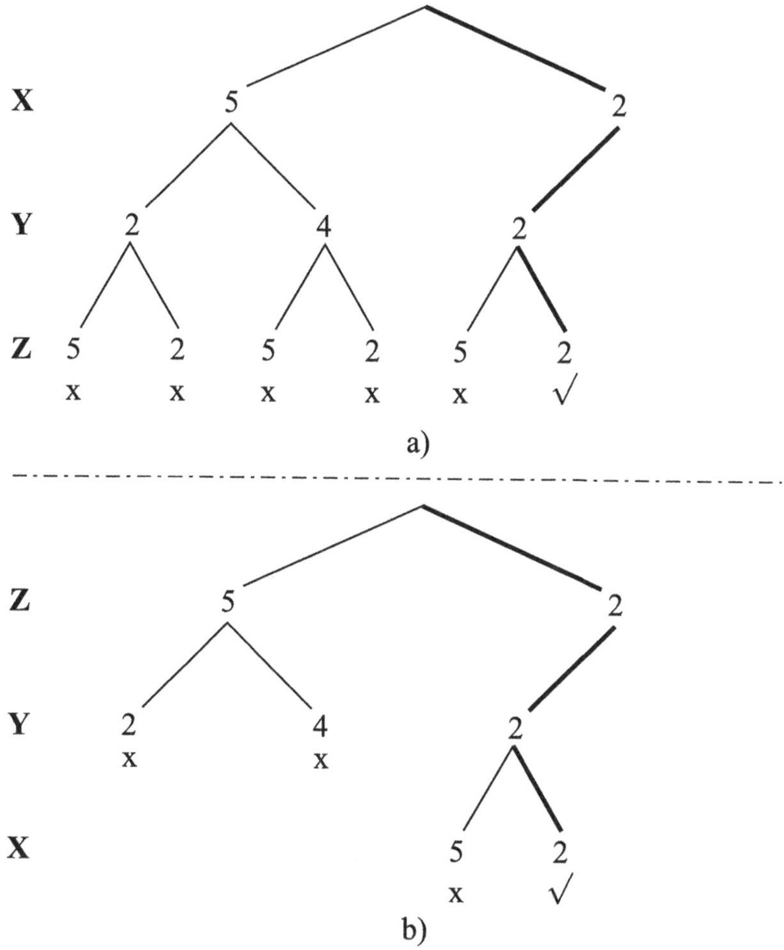

Figure 4.3. *Search trees explored by Bt to seek a single solution for the illustrative example: a) according to the order X, Y and Z; b) according to the order Z, Y and X*

Note, then, that the two trees for each type of search differ in size (the number of nodes is different from one tree to the other). What we can say is that the order of traversal for the variables has a significant influence on the performance of the algorithms in terms of the number of nodes visited and consequently in terms of CPU time.

As before, the size of the search space explored also depends on the order of traversal for the values associated with the variables. In the previous traversal, the values were traversed according to the order of their appearance in the domains. Thus, for the first case and following the ascending order of the values, we will obtain the two trees shown in Figure 4.4.

In the context of this simple example, no difference can be seen with the first case (seeking all of the solutions) except that the processing order for the two main branches of the tree is reversed.

For the second case (seeking the first solution if a solution exists) and following the ascending order of the values, we obtain the two trees shown in Figure 4.5.

This time, we note that the order of value assignment has a considerable effect on the reduction in size of the space explored by the algorithm. Thus, the order of value traversal also has a significant influence on the performance of the algorithms in terms of the number of nodes visited and consequently in terms of CPU time.

From this very simple example, it appears that the notion of order is crucial for solving large CSPs. Indeed, the use of an ideal order would reduce the search space, which would lead to the immediate discovery of a solution to the problem [MIN 93].

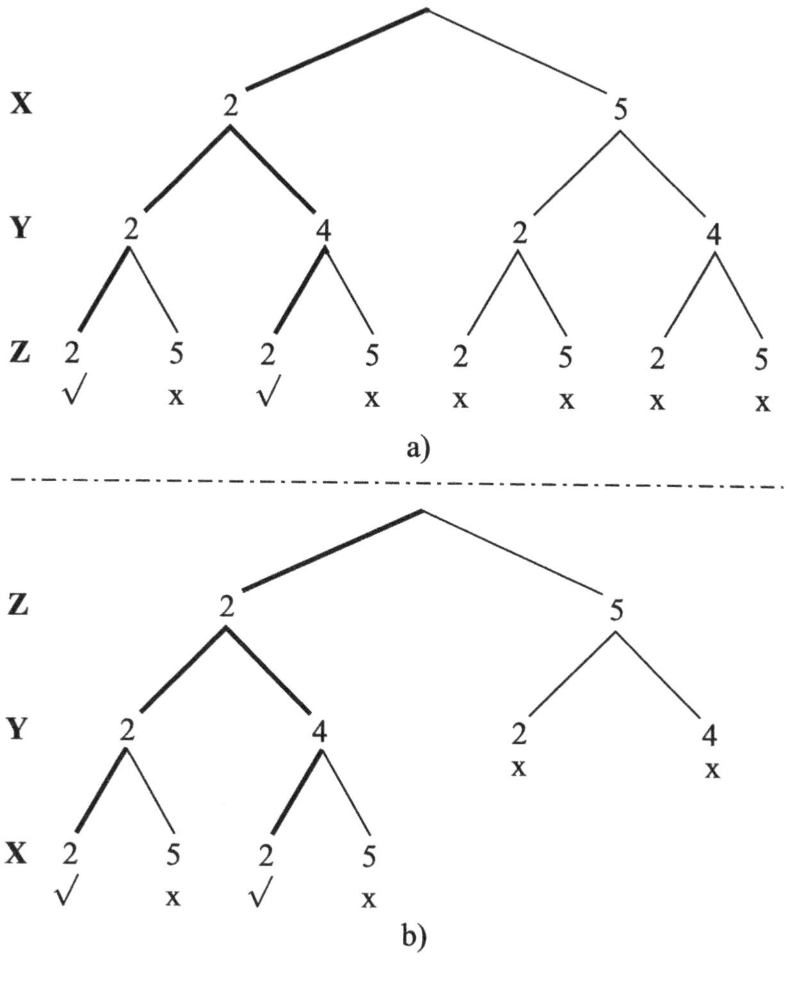

Figure 4.4. *Search trees explored by Bt for seeking all of the solutions to the illustrative example: a) according to the order X, Y, Z and the ascending order of the values; b) according to the order Z, Y, X and the ascending order of the values*

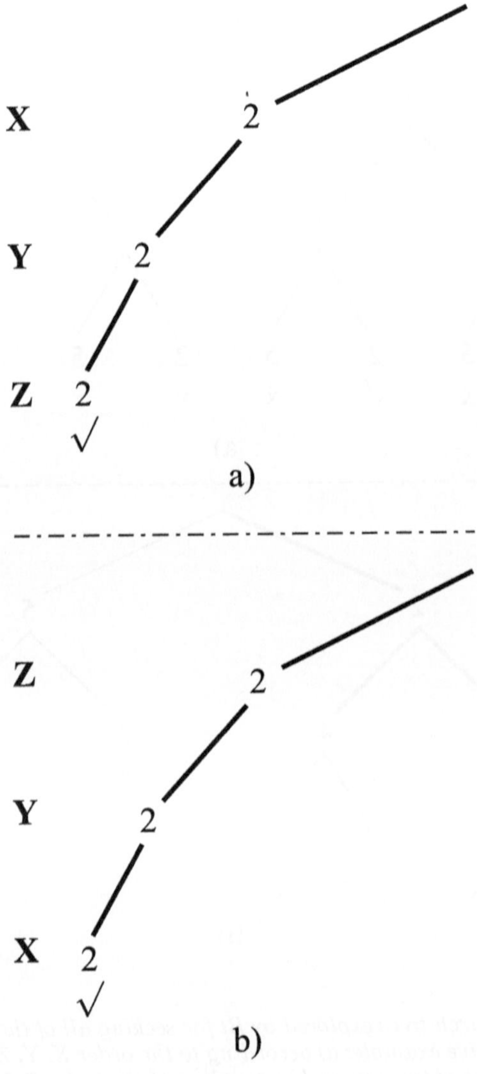

Figure 4.5. *Search trees explored by Bt for seeking a single solution to the illustrative example: a) according to the order X, Y, Z and the ascending order of the values; b) according to the order Z, Y, X and the ascending order of the values*

Furthermore, it is rare, in the CSP community, to seek all the solutions to a problem. Indeed, the majority of the CSP problems aim to produce a single solution. In this case, it is preferable and advisable to orientate the search toward the most promising paths so as to obtain a solution as quickly as possible, hence the importance of heuristics.

In the literature, the following search orders are considered: variable ordering or vertical ordering, value ordering or horizontal ordering and that of the constraint tightness. A combination of these orders was proposed in [KHA 04]. In what follows, we will review the most widespread heuristics.

4.2.2. *Variable ordering*

There are numerous generic heuristics for ordering variables, which may have a significant impact on the efficiency of the search since they decide when the satisfaction of a constraint can be verified and therefore affect the size of the space explored [ALL 94]. These heuristics can be used statically or dynamically:

– Statically, i.e. before execution to construct an order that will be used in the same way in all branches of the search tree.

– Dynamically, i.e. by seeking during the execution of the algorithm and at each node, the next variable to be instantiated. During a search, this dynamic ordering has the advantage of taking better account of the current state of the problem with respect to the current partial instantiation.

4.2.2.1. *Maximum degree heuristic*

Maximum degree (MD) heuristic is a static heuristic that consists of ordering the variables according to the decreasing order of the variable's degrees, knowing that the degree of a variable represents the number of constraints in which it is involved. It takes the list of the CSP's variables as an argument and returns the same list arranged in decreasing order of their degrees.

```
Max-degree (ListVar)
Begin
    i:= | ListVar | /*the cardinality of ListVar*/
    While i ≥ 1 do
      j:= 2
      While j ≤ i do
        If Degree (ListVar[j]) > Degree (ListVar [j-1]) then
          Switch (ListVar [j]), ListVar [j-1]))
        End if
        j:= j + 1
      End while
      i:= i-I
    End while
    Return (ListVar)
End
```

Algorithm 4.1. *The max-degree heuristic*

4.2.2.2. *Maximum tightness heuristic*

Maximum tightness (MT) heuristic is a static heuristic that consists of ordering the variables according to the decreasing value of the tightness's sum of constraints that involve each variable. Note that the tightness of a constraint C_{ik} is equal to the following ratio:

The number of couples prohibited by C_{ik} */ the number of possible couples:*

where *the number of possible couples* $= |D_i| \times |D_k|$.

It takes the list of the CSP's variables as the argument and returns the same list arranged in the decreasing order of the tightness's sum of constraints in which it is involved.

```
Max-Tight (ListVar)
Begin
  i:= |ListVar|/*the cardinality of ListVar*/
    While i ≥ 1 do
      j:= 2
        While j ≤ i do
            If Sum-Tight (ListVar [j]) > Sumj-Tight(ListVar [j-1]) then
                Switch (ListVar[j]), (ListVar [j-1])
            End if
            j:= j + 1
        End while
      i:= i - 1
  End while
  Return (ListVar)
End
```

Algorithm 4.2. *The max-tight heuristic*

4.2.2.3. *Minimum width heuristic*

Minimum width (MW) heuristic is a static heuristic that consists of ordering the variables from last to first by repeatedly selecting a variable in the constraint graph that connects to the minimal number of variables that have not yet been selected [FRO 95].

It takes the list of the CSP's variables as the argument, creates an empty list ListVarInter, creates a copy CopyListCont of the list of constraints, then it seeks the variable with the minimum degree in the graph of constraints, removes it from ListVar and removes all the constraints involving this CopyListCont variable. This variable will be added to the list of variables. When CopyListCont is empty, the remaining variable in ListVar is added to ListVarInter. This heuristic returns the reversed ListVarInter list.

```
Min-Width (ListVar)
Begin
    ListVarInter:= Ø /*an initially empty list*/
    CopyListCont:= a copy of the list of constraints
    While not empty (CopyListCont) do
      x:= Var-Min-Degree (ListVar)
      Delete (ListVaR, x)
      Add (ListVarInter, x)
      i:= 1
        While i ≤ |CopyListCont| do
          y:= CopyListCont[i]
          If includes (y, x) then
            Delete (CopyListCont, y)
          Else
            i:= i+1
          End if
        End while
      End while
      Add (ListVarInter, Last-Remaining-Var in ListVar)
      Return (Reverse (ListVarInter))
End
```

Algorithm 4.3. *The min-width heuristic*

4.2.2.4. *Maximum cardinality heuristic*

Maximum cardinality (MC) heuristic is a dynamic heuristic that consists of ordering the variables according to the following procedure:

```
Max-card (ListVar, ListAssignedVar)
Begin
    max:= 0
    cpt:= 0
    For each v ∈ ListVar do
      For each ass V ∈ ListAssignedVar do
        If v and ass V ∈ the same constraint then
          cpt:= cpt + 1
        End if
      End for
      If max < cpt then
        max:= cpt
        mostvar:= v
      End if
    End for
    Add (ListAssignedVar, mostvar)
    Return (mostvar)
End
```

Algorithm 4.4. *The max-card heuristic*

The first variable is selected randomly then, at each step, the chosen variable will be that which is connected with the greatest number of variables that have already been selected [GEN 96].

4.2.2.5. *First fail principle heuristic*

First fail principle (FFP) heuristic was set out by Haralick and Elliot in 1980. It is based on the following principle: "To succeed, try first where you are most likely to fail" [SMI 96, SMI 97]. It then, first of all, consists of choosing the most constrained variables to highlight the inconsistency of the problem as soon as possible.

4.2.2.6. *Smallest domain heuristic*

Smallest domain (SD) heuristic is a dynamic heuristic due to the fact that the size of the domain changes with the progression of the algorithm's execution.

```
Smallest-Domain (ListVar)

Begin

  max:= 0
  dsize:= 0
  For each v ∈ ListVar do
    dsize:= Domain-Size (v)
    If dsize > max then
      max:= dsize
      sd-var:= v
    Else If dsize = max then
      If degree (v) > degree (sd-var) then
        sd-var:= v
      End if
    End if
  End for
  Return (sd-var)

End
```

Algorithm 4.5. *The smallest domain heuristic*

This heuristic is based on the FFP and is also called the "smallest remaining domain" or the "minimum current domain (MCD)". It consists of choosing the variable whose current domain is of minimum size. In cases where there are multiple variables verifying this condition (tie-break situation), a random variable or a variable selected by a static order (MD, MT, MW, etc.) can be chosen from them. If, during a tie-break, the maximum degree variable is chosen, we speak of BZ-heuristic [BRE 1979]. This heuristic is often recommended and can be used with algorithms such as forward-checking and its extensions since they anticipate the effect of the instantiated variables on those not yet instantiated. They can thus choose a variable that may cause an immediate failure.

4.2.3. *Value ordering*

Once the variable to be instantiated is selected, the order in which its values will be considered can have a very significant impact on the CPU time. As in the case of vertical ordering, we can consider the order of value appearance, the alphanumeric order if possible or even randomly.

In the following, we present heuristics whose search is less blind than that of the heuristics mentioned above. Note that a horizontal heuristic will have no effect on the efficiency of the search if the CSP is inconsistent or if we seek all solutions since the whole width of the tree will then have to be explored.

4.2.3.1. *Look-ahead value ordering*

This is a generic heuristic proposed by [FRO 94, FRO 95]. It consists of counting the number of times that a value of the current variable's domain is in conflict with a value of a future variable. Priority will be given to the value leading to the smallest number of conflicts. As this heuristic is associated with the FC-CBJ hybrid algorithm, it gives good results particularly with over-constrained problems.

4.2.3.2. *Min-conflict heuristic*

This heuristic was proposed by [MIN 92]. It favors the most promising values since this ordering is relatively done according to the "future" of the search consisting of the variables that have not yet been instantiated. This heuristic consists of randomly choosing a variable from those in conflict, i.e. the variables with at least one violated constraint, then seeking the best value for this variable which minimizes the number of constraints violated.

This heuristic has proved to be very efficient and has provided spectacular results, especially for problems having a significant density. This efficiency is due to the fact that it guides the repairs and the search toward promising regions.

4.2.3.3. *Sticking value heuristic*

This heuristic was proposed by [FRE 82]. In a non-chronological look-back strategy, sticking value recalls the value that was assigned to a variable during the forward phase. Then, during a new forward phase it proposes to reinstantiate this variable to its former value if this value also proves to be compatible with the new commitment. The feeling here is that a variable that has once been compatible may be so again especially if we consider that the previous jumps of the new forward phase only generally challenge a small part of the commitment. This heuristic that maintains a log during the search is very close to backmarking. However, its implementation is considerably simpler. The results reported by Frost and Dechter show that incorporating heuristics into backjumping significantly improves the search when the domains remain of a reasonable size.

4.2.4. *Constraints-based ordering*

The choice of the variables to be instantiated can be made according to the order of appearance of the associated constraints or the alphanumeric order of these constraints or even randomly from these constraints. But, the most effective heuristic is that which favors the tightest constraints to detect a failure as soon as possible, this is within the framework of satisfying all constraints.

Remember that the tightness of a binary constraint C is equal to the number of couples forbidden by C divided by the total number of couples possible for C.

The greater the tightness, i.e. the closer to 1, the more difficult it is to satisfy the associated constraint. With tightness equal to 1, all couples are forbidden and therefore the constraint is always violated.

And the weaker it is, i.e. close to 0, the easier it is to satisfy the associated constraint. With zero tightness, all the couples are authorized and the constraint is therefore always satisfied.

4.3. Bibliography

[ALL 94] ALLIOT J.M., SCHIEX T., *Intelligence Artificielle et Informatique Théorique*, 2nd ed., Cépaduès-Edition, 1994.

[ALL 96] ALLEN J., MINTON S., "Selecting the right heuristic algorithm: runtime performance predictors", *Proceedings of the Canadian Artificial Intelligence Conference*, LNCS, Springer, 1996.

[BAC 95] BACCHUS F., VAN RUN P., "Dynamic variable ordering in CSP", *Proceedings of Principles and Practice of Constraint Programming*, LNCS, Springer, pp. 258–275, 1995.

[BRE 1979] BRELAZ D., "New methods to color the vertices of a graph", *Communications of the ACM*, volume 22, number 4, 1979.

[DEC 89] DECHTER R., MEIRI I., "Experimental evaluation of preprocessing techniques in constraint satisfaction problems", *Proceedings of the International Joint Conference on Artificial Intelligence*, Detroit, MI, USA, pp. 271–277, 1989.

[FRE 82] FREUDER E.C., "A sufficient condition for backtrack-free search", *Journal of the ACM*, vol. 29, pp. 24–32, 1982.

[FRO 94] FROST D., DECHTER R., "In search of the best constraint satisfaction search", *Proceedings of the National Conference on Artificial Intelligence*, Seattle, Washington, USA, AAAI Press, pp. 301–306, 1994.

[FRO 95] FROST D., DECHTER R., "Look-ahead value ordering for constraint satisfaction problems", *Proceedings of the International Joint Conference on Artificial Intelligence*, Montreal, Canada, pp.572–578, 1995.

[GEN 96] GENT I.P., MACINTYRE E., PROSSER P., SMITH B.M., WALSH T., "An empirical study of dynamic variable ordering heuristics for the constraint satisfaction problem", *Proceedings of Principle and Practice of Constraint Programming*, Cambridge, MA, USA, pp. 179–193, 1996.

[GHÉ 93] GHÉDIRA K., MASC une approche multi-agents des problèmes de satisfaction de contraintes, PhD Thesis ENSAE, Toulouse, France, 1993.

[HAM 99] HAMADI Y., Traitement des problèmes de satisfaction de contraintes distribués, PhD Thesis, Montpellier II, France, 1999.

[JÉG 91] JÉGOU P., Contribution à l'étude des problèmes de satisfaction de contraintes: algorithmes de propagation et de résolution, propagation des contraintes dans les réseaux dynamiques, PhD Thesis LIRMM-USTL, Montpellier II, 1991.

[KHA 04] KHAYATI N., Benjaafar I., GHÉDIRA K., "Heuristic cooperation to speed up the constraint satisfaction", *Proceedings of the IASTED International Conference on Artificial Intelligence and Soft Computing*, Marbella, Spain, 2004.

[LUI 95] LUI J., SYCARA K.P., "Emergent constraint satisfaction through multi-agent coordinated interaction", *Proceedings of the European Workshop on Modelling Autonomous Agents in a MultiAgent World*, Neuchatel, Switzerland, 1995.

[MIN 92] MINTON S., JOHNSON M.D., PHILIPS A.B., LAIRD P., "Minimizing conflicts: a heuristic method for constraint-satisfaction and scheduling problems", *Artificial Intelligence*, vol. 58, pp. 161–205, 1992. [Reprinted in Constraint-based Reasoning, FREUDER E.C., MACKWORTH A.K. (eds), MIT Press, 1994.]

[MIN 93] MINTON S., "Integrating heuristics for constraint satisfaction problems: a case study", *Proceedings of the National Conference on Artificial Intelligence*, Washington, D.C., USA, 1993.

[MIN 96] MINTON S., "Is there any need for domain dependent control information? A reply", *Proceedings of the National Conference on Artificial Intelligence*, Portland, Oregon, USA, 1996.

[RAO 93] RAO V.N., KUMAR V., "On the efficiency of parallel backtracking", *IEEE Transactions on Parallel and Distributed Systems*", vol. 4, no. 4, pp. 427-437, 1993.

[SMI 96] SMITH B.M., Succeed-first or fail-first: a case study in variable and value ordering, Research Report 96.26, School of Computer Studies, University of Leeds, 1996.

[SMI 97] SMITH B.M., GRANT S.A., Trying harder to fail first, Research Report 97.45, School of Computer Studies, University of Leeds, 1997.

[YOK 92] YOKOO M., DURFEE E.H., ISHIDA T., KUWABARA K., "Distributed constraint satisfaction for formalizing distributed problem solving", *International Conference on Distributed Computing Systems*, Yokoama, Japan. IEEE Computer Society Press, 1992.

Chapter 5

Learning Techniques

The first real learning mechanism for CSPs appeared in 1986 under the definition “learning while searching” [DEC 86]. This mechanism implicitly or explicitly manipulates specific induced constraints, whose generation, recording and reuse do not change the set of CSP solutions and in particular prevent the search from visiting futile regions.

Incorporating refinements such as recording and learning mechanisms into the various techniques cited in Chapter 3 gives them greater efficiency. The task of these mechanisms is to intelligently record the reasons for the dead-end so that the same conflicts will not arise again in the continuation of the search or across instances of the same domain. When the algorithm terminates, the information accumulated by the learning mechanisms is part of a new, more knowledgeable, representation of the same problem. That is, if the algorithm is executed once again on the same input, it will have a better performance [DEC 86, DEC 90].

First, we begin by defining the notions relating to the “nogood” concept. Subsequently, we will present the nogood-recording (NR) algorithm and its extensions.

5.1. The "nogood" concept

DEFINITION 5.1:– A NOGOOD.– *A nogood is a couple (A, J) where A is a partial instantiation of X and J a subset of constraints of C, such that no solution to the CSP (X, J) contains A.*

More explicitly, a nogood is a partial assignment (i.e. an assignment of a subset of variables) that is not part of any solution to the global problem or, similarly, a partial assignment forbidden by certain problem's constraints. This is also referred to as a globally inconsistent partial assignment. Note that any inconsistent partial assignment is a nogood, but the opposite is not necessarily true. This notion of nogood serves as a guide to avoid the useless exploration of inconsistent regions by pruning the search tree. Filtering via local consistency reinforcement is the first method for producing nogoods. It uses the following property:

PROPERTY 5.1:– *Let A be a consistent partial assignment; let x be a variable not assigned by A and dom(x) its domain; A is a nogood if and only if* $\forall v \in dom(x)$, $A \cup \{(x, v)\}$ *is a nogood.*

A second method for producing nogoods was proposed in [DEC 90], then improved in [SCH 93, SCH 94a]. It is based on a systematic backtrack search and uses the following three properties:

PROPERTY 5.2:– BASIC NOGOODS.– *Let A be a partial assignment; let C be a constraint; if C is not satisfied by A, then (A, {C}) is a nogood.*

Nogoods generated in this way provide no new information since the inconsistency of *A* is contained in the constraint *C* itself.

PROPERTY 5.3:- NOGOODS UNION.– *Let A be a partial assignment; let X be a non-instantiated variable; and* $A_1, \ldots, A_m$ *be all possible extensions of A according to X. If* $(A_1, J_1), \ldots, (A_m, J_m)$ *are nogoods, then* (A, J) *where* $J = \cup_i J_i$ *is a nogood.*

This mechanism can be used to construct a new nogood (A', J) from a given nogood (A, J) where $A' \subset A$. This new nogood has a greater pruning power.

PROPERTY 5.4:- PROJECTION OF NOGOODS.– *Let* (A, J) *be a nogood; let* X_j *be the set of variables involved in the set J; and let* $A \downarrow X_j$ *be the projection of A on* X_j*, then* $(A \downarrow X_j, J)$ *is a nogood.*

Note that the constraints' involvement in the nogoods has the dual advantage of allowing for the explanation of inconsistencies (set of responsible constraints) and facilitating the processing of dynamic problems.

Relying on the last three properties (5.2, 5.3 and 5.4), it is possible to produce nogoods during a standard search. During a standard backtrack, the first property can be applied to any failure leaf (unsatisfied completely instantiated constraint), and the two others to any failure node.

5.1.1. *Example of union and projection*

Consider the simple CSP, with three variables, as shown in Figure 5.1. It should be noted that the values taken by these variables are ordered as follows: $a < b < c$.

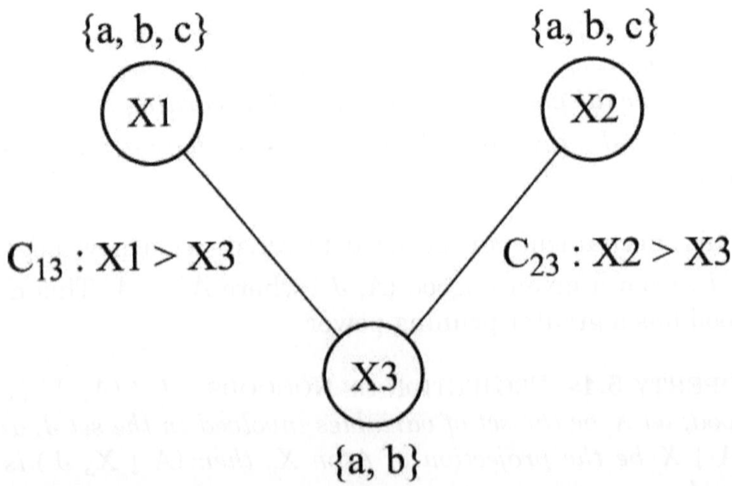

Figure 5.1. *Example of CSP illustrating union and projection*

The instantiation $A_1 = \{(X_1, a), (X_2, a), (X_3, a)\}$ is consistent since it violates the constraint C_{13}, hence $(A_1, \{C_{13}\})$ is a nogood.

The instantiation $A_2 = \{(X_1, a), (X_2, a), (X_3, b)\}$ violates the same constraint C_{13}. We can then deduce that $(A_2, \{C_{13}\})$ is a nogood.

None of these nogoods will be incorporated into the CSP since the information that they provide is expressed by the constraint C_{13} itself.

We can now apply the union and thus infer that: $(\{(X_1, a), (X_2, a)\}, \{C_{13}\})$ is a nogood. By applying the projection, we will get the nogood $(\{(X_1, a)\}, \{C_{13}\})$ since only the variable X_1 is involved in the constraint C_{13}.

It should be noted that this nogood has the greatest pruning power if the CSP is more significant.

5.1.2. *Use of nogoods*

A generated nogood can be used to carry out the following three different tasks:

1) Backjumping

Each time a nogood (A, C) is produced at a failure node by union and projection, it is possible to perform a direct backtrack to the most recently instantiated variable of A. This possibility comes directly from the fact that if (A, C) is nogood, (A, C') is a nogood for each $A' \subset A$. It allows numerous unnecessary backtracks to be avoided and to go directly toward the source of the failure.

2) Reduction of the current search

The nogoods produced during the search are induced constraints that are added to the set of CSP's constraints. Such a task has the advantage of avoiding certain useless branches of the search tree.

3) Reduction of the search for close problems

A nogood (A, C) produced within the framework of a CSP P resolution is obviously valid within this framework and more generally within the framework of any CSP P' whose set of constraints contains C. This allows nogoods to be produced during the previous resolutions and to be used to solve the current CSP, with the same advantage of avoiding unpromising subspaces.

5.1.3. *Nogood handling*

Although the recording and learning mechanism is a very useful tool for solving CSPs, the number of nogoods increases exponentially with the size of the problem [DEC 90].

Therefore, it is wise to limit storage to a reasonable subset of nogoods. Hence, the need for a policy to produce and retain only the most interesting nogoods, i.e. those with the greatest pruning power. For example, it is possible to limit the arity of learned constraints: to only produce nogoods (A, C) such that the cardinality of A is less than or equal to an arbitrary constant K. This is called K-th-order learning [DEC 90]. Such a task limits the number of nogoods recorded, their size and the time needed for their detection.

Among the hybrid algorithms belonging to this category, we cite the following: NR [SCH 93, SCH 94a], NR forward-checking [SCH 93, SCH 94a], stubborn NR [SCH 94b], weak-commitment NR [RIC 95b] and weak-commitment NR forward-checking [RIC 95b].

5.2. Nogood-recording algorithm

An algorithm, stemming from the "learning while searching" approach [DEC 86], is proposed in [SCH 93, SCH 94a] under the name of NR. It relies on the recording of nogoods. Within the framework of static problems, these nogoods can be used to prune the search space. Within the framework of dynamic problems, the fact that the nogoods are created with their justifications (a subset of explicit constraints for the problem from which they result) allows, on the one hand, those whose justification is included in these current explicit constraints to be reused and, on the other hand, the current search space to be pruned.

NR improved the shallow learning algorithm. Intelligent backtracking carried out by NR corresponds to that carried out by the conflict-directed-backjumping (CBJ) algorithm introduced in [PRO 93], which means that this algorithm is a hybridization mixing chronological backtracking and constraint recording.

NR is a recursive algorithm that takes an initially empty instantiation I and a sequence V of variables to be instantiated as argument, and which returns a set of constraints J denoted as conflict set. As defined, NR calls upon various functions to determine a solution:

– The *check* (I) function checks whether the instantiation I is consistent. If so, an empty set is returned. If not, there is at least one violated constraint. So, if a violated constraint c was recorded following *a nogood* (I, J), then the function returns the set J, justification for the nogood. Otherwise, c is returned.

– The *record* function adds the induced constraints corresponding to the constructed nogoods to the CSP.

– The *project* (I, J) function projects the instantiation I on the set of variables involved in the constraints of J.

– The *involved* (x, J) function checks whether the variable x is involved in constraints of the set J.

NR is a natural extension of backtracking. Besides, various other algorithms hybridizing tree search and learning have been extended and developed, we include: value-based learning [FRO 95], graph-based shallow learning [FRO 95], jump-back learning [FRO 95], deep learning [FRO 95] and generalized nogood learning [FRE 94]. In the following paragraphs, we show how NR works on the example presented in section 5.1.1.

Iteration	NR(I,V)	J	BJ	I'	K	Return
1	NR({},{X_1, X_2, X_3})	ϕ	false	{(X_1,a)}	ϕ	
2	NR({(X_1,a)},{X_2,X_3})	ϕ	false	{(X_1,a),(X_2,a)}	ϕ	{C_{13}} ↑
3	NR({(X_1,a),(X_2,a)},{X_3})	ϕ	false	{(X_1,a),(X_2,a),(X_3,a)}	{C_{13}}	
		{C_{13}}		{(X_1,a),(X_2,a),(X_3,b)}	{C_{13}}	
		{C_{13}}				{C_{13}}

```
NR (I, V)
Begin
   If V = Ø then
     I is a solution
   Else
     Let x ∈ V, J = Ø, BJ = False
     For each v ∈ Dx To BJ do
       Let I':= I ∪ {(x, v)}, K:= Check (I')
       If K = Ø then
         J-sons:= NR (I', V\{x})
         If involved (x, J-sons) then
           J:= J ∪ J-sons
         Else
           J:= J-sons
           BJ:= True
         End if
       Else
         J:= J ∪ K
       End if
     End for
     If BJ = False then
       record (project (I, J), J)
     End if
     Return(J)
   End if
End
```

Algorithm 5.1. *Nogood-recording*

After applying union and projection (see section 5.1.1) to the two found Nogoods $\{\{(X_1, a), (X_2, a), (X_3, a)\}, \{C_{13}\}\}$ and $\{\{(X_1, a), (X_2, a), (X_3, b)\}, \{C_{13}\}\}$, we obtain:

– the new following nogood: $\{\{(X_1, a)\}, \{C_{13}\}\}$;

– the third iteration returns $\{C_{13}\}$;

– the variable X_2 is not involved in C_{13}; therefore, there is once again a return of C_{13} by the second use of NR;

– the variable X_1 is involved in C_{13};

– $\{\{(X_1, b), (X_2, a)\}, \{C_{23}, C_{13}\}\}$ is a nogood;

– the couple $\{C_{23}, C_{13}\}$ is returned;

– the variable X_2 is involved in C_{23}.

Iteration	NR(I,V)	J	BJ	I'	K	Return
3'		$\{C_{13}\}$		$\{(X_1,b)\}$	ϕ	↑
4	$NR(\{(X_1,b)\},\{X_2,X_3\})$	ϕ	false	$\{(X_1,b),(X_2,a)\}$	ϕ	
5	$NR(\{(X_1,b),(X_2,a)\},\{X_3\})$	ϕ	false	$\{(X_1,b),(X_2,a),(X_3,a)\}$	$\{C_{23}\}$	
		$\{C_{23}\}$		$\{(X_1,b),(X_2,a),(X_3,b)\}$	$\{C_{13}\}$	
		$\{C_{23}, C_{13}\}$				$\{C_{23}, C_{13}\}$

Iteration	NR(I,V)	J	BJ	I'	K	Return
5'		$\{C_{23}, C_{13}\}$		$\{(X_1,b), (X_2,b)\}$	ϕ	
6	$NR(\{(X_1,b),(X_2,b)\},\{X_3\})$	ϕ	false	$\{(X_1,b),(X_2,b),(X_3,a)\}$	ϕ	
7	$NR(\{(X_1,b),(X_2,b),(X_3,a\},\{\})$	$\{(X_1,b),(X_2,b),(X_3,a\}\}$ is a solution				

5.3. The nogood-recording-forward-checking algorithm

This algorithm is based on the same principle as NR, except that it replaces the *check* function with the *forward-check* function and adds an *unforward* function.

1) The *forward-check* function establishes *forward-checking* and at each step records the constraints that have led to the reduction of a domain D_i into value killers of the variable X_i. If a domain of a variable becomes empty, the process of *forward-checking* stops and the set of *value killers* are returned; otherwise the function returns an empty set.

2) The *unforward* function restores the modified domains and the set of *value killers* to their previous states.

This algorithm records appreciably less nogoods than NR for a given learning order. This is not surprising: filtering carried out at each node by FC limits the possibilities of failure and therefore of learning.

Table 5.4 illustrates the resolution phases for the CSP given in section 5.1.1.

Iteration	NR-FC(I,V)	J	BJ	I'	K
1	NR-FC({},{X_1,X_2,X_3})	φ	False	{(X_1,a)}	φ
2	NR-FC({(X_1,a)},{X_2,X_3})	φ	False	{(X_1,a),(X_2,a)}	{C_{13}}
3	NR-FC({(X_1,b)},{X_2,X_3})	φ	False	{(X_1,b),(X_2,a)}	{C_{23}}
		{C_{23}}		{(X_1,b),(X_2,b)}	φ
4	NR-FC({(X_1,b),(X_2,b)},{X_3})	φ	False	{(X_1,b),(X_2,b) ,(X_3,a)}	φ
5	NR-FC({(X_1,b),(X_2,b) ,(X_3,a},{})	{(X_1,b),(X_2,b) ,(X_3,a}} is a solution			

This extension was introduced in [RIC 95b]. Compared with weak-commitment, the difference resides in the *assign-NR* (*x*, *VarsDone*, *VarsLeft*, *D*, *C*) function. For the weak-commitment algorithm, inconsistent-vals (x, D_x, VarsDone, C) returns $IncD_x$; the set of values of x inconsistent with the partial instantiation *VarsDone*, while for WC–NR it returns ($IncD_x$, *NG-set*) where *NG-set* is relative to the set of nogoods that forbids x any value in $IncD_x$. The prohibition of all values leads to the production of a new nogood via union. The *assign-NR* process conforms to the following procedure:

Note that this algorithm served as the basis to develop other hybrid algorithms [RIC 95b] and that learning in the context of a non-systematic search, i.e. via iterative repair, has also sparked the interest of several other researchers [JIA 94, RIC 95a].

```
NR-FC(I, V)
Begin
  If V = Ø then
    I is a solution
  Else
    Let x ∈ V, J = Ø, BJ = False
      For each v ∈ D_x To BJ do
        Let I':= I ∪ {(x, v)}, K:= Forward-Check (I', x)
        If K = Ø then
          J-sons:= NR-FC (I', V\{x})
          un-forward (x)
          If involved (x, J-sons) then
            J:= J ∪ J-sons
          Else
            J:= J-sons
            BJ:= True
          End if
        Else
          J:= J ∪ K
          un-forward (x)
          record(project(I, K), K)
        End if
      End for
      If BJ = False then
        record (project (I, J), J)
      End if
        Return(J)
  End if
End
```

Algorithm 5.2. *NR–FC*

5.4. The weak-commitment-nogood-recording algorithm

```
WC-NR (VarsDone, VarsLeft, D, C)
Begin
    If VarsLeft =Ø then
      Return (VarsDone)
    Else if (VarsDone ∪ VarsLeft) ⊨ C then
      Return (VarsDone ∪ VarsLeft)
    Else
      Let x ∈ VarsLeft involved on a constraint violation
      VarsLeft:= VarsLeft\ {x}
        Result:= Assign-NR (x, VarsLeft, VarsDone, D, C)
      If Result = True then
          WC-NR (VarsDone, VarsLeft, D, C)
      Else
          Nogood-Record (VarsDone)
          VarsLeft:= VarsLeft ∪ VarsDone
          VarsDone:= Ø
          WC-NR (VarsDone, VarsLeft, D, C)
      End if
    End if
End
```

Algorithm 5.3. *WC–NR*

```
Assign-NR (x, VarsDone, VarsLeft, D, C)
Begin
    (IncD, NG-set):= Inconsistent-Vals-NG (x, D_x, VarsDone, C)
    ConD:= D_x – IncD
    If ConD = Ø then
      Result:= Merge (NG-set, x) /* Production of new
                                    Nogoods using the union
      Return (Result)
    Else
      v:= Min-Conflict-Val (x, ConD, VarsDone, VarsLeft)
      VarsDone:= VarsDone {(x, v)}
      Return (True)
    End if

End
```

Algorithm 5.4. *The assign-NR procedure*

5.5. Bibliography

[DEC 86] DECHTER R., "Learning while searching in constraint satisfaction problems", *Proceedings of the National Conference on Artificial Intelligence*, Philadelphia, USA, AAAI Press, pp. 178–183, 1986.

[DEC 90] DECHTER R., "Enhancement schemes for constraint processing: backjumping, learning and cutset decomposition", *Artificial Intelligence*, vol. 41, pp. 273–312, 1990.

[FRE 94] FREUDER E.C., WALLACE R.J., "Generalizing inconsistency learning for constraint satisfaction", *Proceedings of the European Conference on Artificial Intelligence*, Amsterdam, The Netherlands, 1994.

[FRO 95] FROST D., DECHTER R., "Look-ahead value ordering for constraint satisfaction problems", *Proceedings of the International Joint Conference on Artificial Intelligence*, Montreal, Canada, pp. 572–578, 1995.

[JIA 94] JIANG Y., RICHARDS E.T., RICHARDS B., "Nogood backmarking with min-conflict repair in constraint satisfaction and optimization", *Proceedings of Principles and Practice of Constraint Programming*, LNCS, Springer, pp. 21-39, 1994.

[PRO 93] PROSSER P., "Hybrid algorithms for the constraint satisfaction problems", *Computational Intelligence*, vol. 9, pp. 268–299, 1993.

[RIC 95a] RICHARDS E.T., JIANG Y., RICHARDS B., "NG-backmarking an algorithm for constraint satisfaction", *AIP Techniques for Resource Scheduling and Planning, Bt Technology Journal*, vol. 13, no. 1, January 1995.

[RIC 95b] RICHARDS E.T., RICHARDS B., *Nogood Learning for Constraint Satisfaction*, IC Parc, Imperial College, London, 1995.

[SCH 93] SCHIEX T., VERFAILLIE G., "Nogood-recording for static and dynamic CSP", *Proceedings of the International Conference on Tools with Artificial Intelligence*, Boston, MA, pp. 48–55, 1993.

[SCH 94a] SCHIEX T., VERFAILLIE G., "Nogood-recording for static and dynamic CSP", *International Journal of Artificial Intelligence Tools*, vol. 3, no. 2, pp. 187–207, 1994.

[SCH 94b] SCHIEX T., VERFAILLIE G., "Stubbornness: a possible enhancement for backjumping and nogood recording", *Proceedings of Eurpean Conference on Artificial Intelligence*, Amsterdam, The Netherlands, pp. 165–169, 1994.

Chapter 6

Maximal Constraint Satisfaction Problems

The constraint satisfaction problem (CSP) formalism allows a problem to be represented as a fixed network with constraints to be satisfied at any cost. Unfortunately, there are many problems that are difficult to solve (exponential complexity) or that have no solution that can satisfy all the constraints. In this case, we turn to the maximal constraint satisfaction problems, denoted by max-CSPs, whose aim is to satisfy the maximum of constraints. Several resolution techniques have been adapted from the original CSP framework or have been inspired by methods dedicated to combinatorial optimization. In this chapter, we discuss the following methods:

– the branch and bound and partial forward-checking (PFC) algorithms based on backtracking;

– the weak-commitment algorithm and the GENET method that are essentially based on the min-conflict heuristic;

– the distributed simulated annealing as well as the distributed guided genetic algorithms based on the underlying metaheuristics [GHÉ 07] hybridized with multiagent systems.

6.1. Branch and bound algorithm

The version proposed in [WAL 96] performs a depth search, the *cost* function being the number of violated constraints to be minimized. An upper bound (UB) is set to the maximal number of constraints authorized to be violated (generally equal to the total number of constraints).

Branch & Bound (X, D, C)

Begin

1. ub:= total number of constraints
2. nbcv:= 0
3. I:= Ø
4. Solution:= B & B (X, I, nbcv)

End

Algorithm 6.1. *The branch and bound algorithm*

As soon as the algorithm encounters a feasible solution (leaf in the search tree), the UB will be updated and will be equal to the number of constraints violated by this solution. It should be noted that any current partial instantiation that violates more constraints than the current value of the UB will be ignored along with all of its descendants. Indeed, the corresponding node and its entire subtree are pruned. Then, the algorithm will perform a backtrack on the last instantiated variable when all possible values of the current variable have been tried.

```
B & B (X, I, nbcv)

Begin
    If X = Ø then
      ub:= nbcv
      Return (I)
    Else
      Let x ∈ X
      For each v ∈ Dx do
        nbcvp:= verif-coherence (x, v. I)
        If nbcv + nbcvp ≤ ub then
          B & B (X –m {x}, I ∪ {(x, v)}, nbcv + nbcvp)
        End if
      End for
    End if
End
```

Algorithm 6.2. *The B & B function*

Selection of the initial bound is, nonetheless, crucial. A high bound guarantees that a global optimal solution will be found but proves costly in search time, while a low bound does not require much search time but does not guarantee that a global optimal solution will be found. The difficulty is then to find a bound that offers the best compromise between complexity and quality. Various studies within this framework have been proposed in the literature.

Figure 6.1 shows the application of the branch and bound algorithm to the 4-queens problem. On the left, we represent the numbers of the columns where the queens are placed as well as the columns authorized for the queens that have not yet been placed. On the right, we illustrate this on the chessboard. The crosses represent failures and therefore backtracks.

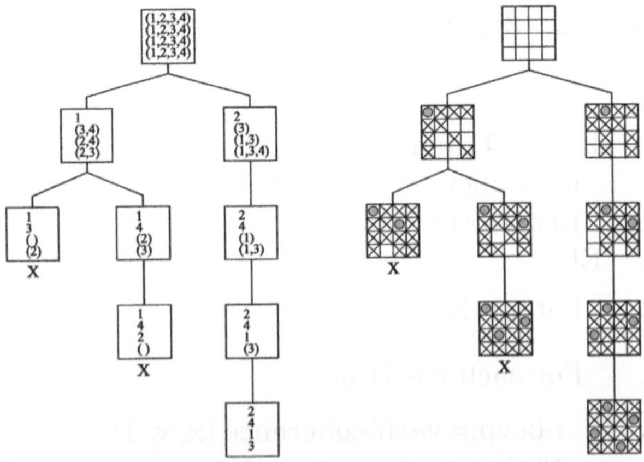

Figure 6.1. *Resolution of the 4-queens problem via branch and bound*

6.2. Partial Forward-Checking algorithm

The forward-checking algorithm for max-CSPs, called PFC, is based on prospection just like the ordinary FC for CSPs. The difference lies in the definition of failure, which is determined as follows: If there is inconsistency, a value is only rejected if the total number of values inconsistent with this value added to the number of violated constraints for the current instantiation becomes greater than or equal to the value of the UB.

This means that the algorithm must dynamically record the number of times a value v has been found to be inconsistent with the currently instantiated values. This number is designated by the *inconsistency count*. It is specific to the value v.

Therefore, the operating principle of the PFC algorithm for the resolution of max-CSPs can be summarized as follows: when a value v is chosen, its own inconsistency count is added to the current number of violated constraints. If the cumulative number obtained is greater than or equal to the value of the UB, then there is a failure.

Thus, for each inconsistent value $val_{i,j}$, instead of removing it as in forward-checking, it increments the new parameter denoted by I_{Cij} (inconsistency count associated with $val_{i,j}$) and then tests whether the number of violated constraints added to this I_{Cij} is smaller than the UB. If so, $val_{i,j}$ is kept, otherwise it is removed. If there is at least one empty domain, then it backtracks as in forward-checking. The backtrack is performed toward the previous variable. This test enables us to estimate whether the current partial instantiation can be extended to a complete instantiation for the improvement of the solution found to date. If a complete instantiation is found, then the UB is updated with the number of corresponding violated constraints, as in the case for branch and bound.

As part of improving the PFC, the *extended forward-checking* (EFC) algorithm was proposed to calculate an estimation of a lower bound for the number of constraints that will be violated when the current partial instantiation is extended to a complete instantiation. Indeed, this algorithm, as described above, assigns an inconsistency-count to a value v based on the inconsistencies that would be incurred by adding v to the current instantiation, that is the choices that have already been made. EFC goes further by forming a lower-bound estimate of the further inconsistencies that would occur in the course of choosing values for each of the remaining variables. For each of these variables, it finds the minimum inconsistency count assigned to the values for that variable. It then adds all of these counts to the inconsistencies incurred in choosing the value $val_{i,j}$. This sum will be a lower-bound estimate of the number of constraints violated by a maximal solution that contains the current constraint and the value $val_{i,j}$. This estimate must obviously be less than the UB.

In what follows, we give the different steps covered by the EFC algorithm to solve the 4-queens problem.

To do this, we use the following notations:

– D_i: all possible squares on line i for queen i;

– *δcurrent*: the current path;

– *nbcv*: the number of constraints violated;

– *δbest:* the best visited path;,

– *Best-cost*: the number of constraints violated for the best visited path to date.

Iteration	D_i	*δcurrent*	*nbcv*	*δbest*	*Best-cost*
1	D_1 = {1,2, 3, 4} D_2 = {1, 2, 3, 4} D_3 = {1, 2, 3, 4} D_4 = {1, 2, 3, 4}	{φ}	0	{φ}	+∞
2	D_1 = {2, 3, 4} D_2 = {1, 2, 3, 4} D_3 = {1, 2, 3, 4} D_4 = {1, 2, 3, 4}	{1}	0	{φ}	+∞
3	D_1 = {2, 3, 4} D_2 = {1, 2, 4} D_3 = {1, 2, 3, 4} D_4 = {1, 2, 3, 4}	{1, 3}	0	{φ}	+∞
4	D_1 = {2, 3, 4} D_2 = {1, 2, 4} D_3 = {2, 3, 4} D_4 = {1, 2, 3, 4}	{1, 3, 1}	1	{1, 3, 1, 2}	2
5	D_1 = {2, 3, 4} D_2 = {1, 2, 4} D_3 = {2, 3, 4} D_4 = {2, 3, 4}	{1, 3}	0	{1, 3, 1, 2}	2
6	D_1 = {2, 3, 4} D_2 = {1, 2, 4} D_3 = {3, 4} D_4 = { 2, 3, 4}	{1, 3}	0	{1,3,1,2}	2

7	$D_1 = \{2, 3, 4\}$ $D_2 = \{1, 2, 4\}$ $D_3 = \{3\}$ $D_4 = \{2\}$	{1, 3, 4}	1	{1, 3, 4, 2}	1
8	$D_1 = \{2, 3, 4\}$ $D_2 = \{1, 2, 4\}$ $D_3 = \{1,2, 3, 4\}$ $D_4 = \{1, 2, 3, 4\}$	{1}	0	{1, 3, 4, 2}	1
9	$D_1 = \{2, 3, 4\}$ $D_2 = \{1, 2\}$ $D_3 = \{2\}$ $D_4 = \{3\}$	{1,4}	0	{1, 3, 4, 2}	1
10	$D_1 = \{2, 3, 4\}$ $D_2 = \{1, 2\}$ $D_3 = \{1, 2, 3, 4\}$ $D_4 = \{1, 2, 3, 4\}$	{1}	0	{1, 3, 4, 2}	1
11	$D_1 = \{2, 3, 4\}$ $D_2 = \{1, 2, 3, 4\}$ $D_3 = \{1, 2, 3, 4\}$ $D_4 = \{1, 2, 3, 4\}$	{φ}	0	{1,3,4,2}	1
12	$D_1 = \{3, 4\}$ $D_2 = \{4\}$ $D_3 = \{1, 3\}$ $D_4 = \{1, 3, 4\}$	{2}	0	{1,3,4,2}	1
13	$D_1 = \{3, 4\}$ $D_2 = \{\}$ $D_3 = \{1\}$ $D_4 = \{1\ 3\}$	{2 ,4}	0	{1, 3 ,4, 2}	1
14	$D_1 = \{3, 4\}$ $D_2 = \{\}$ $D_3 = \{1\}$ $D_4 = \{3\}$	{2, 4, 1}	0	{2, 4, 1, 3}	0

6.3. Weak-commitment search

The weak-commitment algorithm, which was developed by [YOK 94] and is presented below, is based on the min-conflict heuristic [MIN 92]. This algorithm aims to overcome the disadvantages of the tree search techniques and those of the iterative repairs in the sense that it revises a bad instantiation without resorting to an exhaustive traversal.

```
Weak-Commitment (VarsDone, VarsLeft, D, C)

Begin
    If VarsLeft = ∅ then
      Return (VarsDone)
    Else If VarsDone ∪ VarsLeft satisfy all constraints then
      Return (VarsDone ∪ VarsLeft)
    Else
      Let x ∈ VarsLeft involved in a constraint of violation
      VarsLeft:= VarsLeft – {x}
      Result:= Assign (x, VarsLeft, VarsDone, D, C)
      If Result = True then
        Weak-Commitment (VarsDone, VarsLeft, D, C)
      Else
        Nogood-Record (VarsDone)
        VarsLeft:= VarsLeft ∪ VarsDone
        VarsDone:= ∅
        Weak-Commitment (VarsDone, VarsLeft, D, C)
      End if
    End if
End
```

Algorithm 6.3. *The weak-commitment algorithm*

The main idea of weak-commitment is simple. It is primarily based on two lists as well as a mechanism as follows:

– *VarsLeft* references an initial complete instantiation that may possibly be inconsistent.

– *VarsDone* initially empty preserves a consistent partial instantiation during the search.

– A mechanism displacing the variables from the first list to the second list.

With the progress of the algorithm, a variable X_i from VarsLeft involved in the violation of a constraint is supplied to the *assign* procedure to be reinstantiated by a value. This value is chosen so as to be consistent with the partial instantiation preserved by *VarsDone* and to minimize the number of conflicts for the complete instantiation *VarsDone* ∪ *VarsLeft*. The absence of such a value leads to the use of the *nogood-record* procedure, which records *VarsDone* as a *nogood* (definition 5.1). This involves abandoning the partial solution and transforming it into a new constraint in order to avoid constructing a new identical partial solution. Knowing that D is the set of domains and C is the set of CSP's constraints, *assign* conforms to the following procedure:

```
Assign (x, VarsDone, VarsLeft, D, C)

Begin
    IncD:= Inconsistent-Vals (x, D_x, VarsDone, C)
    ConD:= D_x – IncD
    If ConD = Ø then
      Return (VarsDone)
    Else
      v:= Min-Conflict-Val (x, ConD, VarsDone, VarsLeft)
      VarsDone:= VarsDone ∪ {(x, v)}
      Return (True)
    End if

End
```

Algorithm 6.4. *The assign procedure*

6.4. GENET method

The idea behind this method is to use a local search procedure to explore the search space. When the local search is trapped into a local optimum, the cost function is modified to help the search procedure escape it. This method, called GENET for GEneric Neural NETwork, is a repair method based on the min-conflict heuristic [TSA 93]. The basic idea of GENET is to represent a CSP with a neural network (see Figure 6.2). Each value of each variable is represented by a node in the network; such nodes are called instantiation nodes. A state (*on* or *off*) is associated with each instantiation node and those related to the same variable are grouped into a single cluster. During initialization and for each cluster, a single instantiation node is randomly selected to be placed in the *on* state. Each cluster has a modulator that ensures that only one of its instantiation nodes is in the *on* state. A connection is established between each pair of nodes that represent two incompatible instantiations between two different variables. A weight is associated with each connection, which is a negative value that reflects the strength of repulsion that exists between the two connected nodes. At the beginning, all of the connection weights are set to −1.

Each instantiation node, in the *on* state, sends a value (output) to all the other nodes adjacent to it (joined by a connection). This value is multiplied by the weight associated with the connection. An instantiation node in the *off* state produces no *output*. The *input* of an instantiation node X, designated by I_x, is calculated as follows:

$$I_x = \sum W_{x,y} \times S_y;\ such\ that\ X \neq Y$$

where $W_{x,y}$ is the weight associated with the connection between X and Y, and S_y is the state of Y that may be 1 if Y is in the *on* state and 0 if Y is in the *off* state.

Each cluster selects the instantiation node with the greatest input to be in the *on* state. If all the weights are negative, this means we have to choose the instantiation that violates the fewest constraints.

A local minimum can be detected when the network's state does not change and when some nodes receive negative inputs. To move away from local minima, the binary GENET model applies a reinforcement mechanism based on learning. It is a matter of subtracting one of the weights associated with the connections between the *on* nodes. This occurs when the network converges to a local minimum. An energy E of a particular network state is defined by the following formula:

$$Eb = -1/2 \sum\sum W_{i,j} \times S_i \times S_j$$

where i and j represent all the variables. Thus, a solution to the CSP (a global optimum) is represented by a state that has energy equal to 0.

Compared to the conflict minimization heuristic that randomly destroys connections, binary GENET enables an instantiation node – which is currently in the *on* state – to retain its state. The objective is to achieve a network convergence. It has been proved that through learning, the energy in a network will decrease monotonically and so the network will definitely converge.

It should be noted that GENET runs on cycles that are called convergence cycles. During each cycle, each cluster examines its current state and changes one of the instantiation nodes' state to "on" if needed and does so before the next cycle begins.

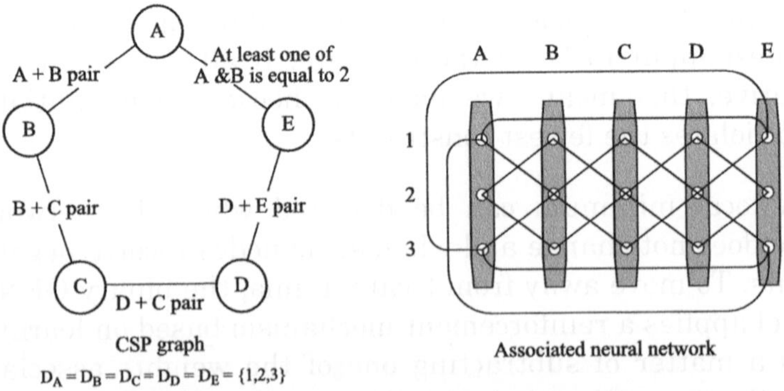

Figure 6.2. *Representation of a binary CSP by GENET*

6.5. Distributed simulated annealing

Distributed simulated annealing was initially proposed in [GHÉ 93] and then readdressed in [GHÉ 94, GHÉ 96]. It aims to maximize the constraint satisfaction on the basis of multi-agent systems. It operates on complete or partial (even empty) instantiations via local repairs guided by the min-conflict heuristic.

The basic model involves two types of agents: the constraint agents and the ariable agents. These agents interact by sending messages; each of them seeks to achieve its maximal satisfaction as follows:

– A constraint is satisfied if it is instantiated and if its relation is respected. Otherwise, it solicits one of its variables and asks it to change value. A constraint is said to be instantiated if all the variables that it involves – also called "its variables" – are instantiated.

– A variable is satisfied if it is instantiated or not solicited (to change value) by one of its constraints, that is in which it is involved. Otherwise, it chooses a new value via the min-conflict heuristic, applies its simulated annealing, and decides to take this new value or keep the old value. It is all the more satisfied, if the number of its satisfied constraints is higher.

The originality of this approach lies, on the one hand, in the distribution of the standard simulated annealing [GHÉ 07] at the level of the variable agents and, on the other hand, in controlling the *temperature* parameter at the level of each variable. Indeed, each variable agent has its own simulated annealing for which it controls all of the parameters and does so on the basis of its own knowledge and objectives. Therefore, it sets its own initial temperature and its evolution over time.

It should be noted that to ensure the termination of this distributed process, the following mechanism was added:

– At zero temperature, a variable only accepts the changes (the values) that lead to a strict increase in local satisfaction expressed by the number of its satisfied constraints. Any value refused, at this temperature level, becomes forbidden for this variable.

– If all its values, except the current instantiation value, are forbidden, then a variable becomes locked.

– An unsatisfied constraint becomes locked when all its variables are locked and, in this case, it no longer seeks to be satisfied.

– The whole process terminates when all the constraints are either satisfied or locked.

It should be noted that this method was successfully applied in different domains, compared to centralized simulated annealing [GHÉ 94, GHÉ 96].

6.6. Distributed and guided genetic algorithm

The distributed guided genetic algorithm, DG^{2}A, is a genetic algorithm guided by the min-conflict heuristic and distributed according to the ideas of ecological niches combined with multiagent systems. The core of this algorithm was proposed for the first time by [GHÉ 02].

It is essentially based on the theory of ecological niches and species [DAR 59]. In nature, living creatures are divided into numerous species. Within the same species, many specimens belong to one distinct sex and both must reproduce to perpetuate the species. It is clear that sexual difference divides a species into two cooperative groups. This bifurcation allows males and females to specialize somewhat, thus covering the range of behaviors required for survival in a broader sense, compared to the possibilities of a single population in competition. Goldberg [GOL 94] illustrated the benefits of cooperation and specialization resulting from natural sexual differences through the maximization of species survival. An ecological niche is represented as the function or role of an organism in a given environment and a species can then be considered as being a class of organisms sharing some characteristics.

In what follows, the basic principles of this algorithm as well as the multi-agent model, the genetic process, and the underlying heuristics will be detailed and illustrated through the 4-queens example. Extensions and improvements of DG^2A will also be explained.

6.6.1. *Basic principles*

By associating a chromosome (individual) with each CSP solution, a gene with each variable and an allele with each value of a variable, [GHÉ 02] were inspired by the work of [TSA 98, TSA 99] to add a new data structure, which would be attached to each chromosome and which we call a "template". They describe the template as consisting of weights called $\delta_{p,q}$, where $\delta_{p,q}$ represents the number of constraints violated by the $gene_{p,q}$ and where p refers to the chromosome and q refers to the position in the chromosome.

For each given chromosome, the so-called penalty operator begins by setting all of the corresponding template's weights to zero and then updates them, as it goes along, according to the status of the constraints as follows: each time a constraint C_j is violated, all the weights related to the genes (variables) that it involves are incremented by 1.

The crossover and mutation operators will attempt to remove the weakest genes, that is the least beneficial, using the min-conflict heuristic.

The templates therefore offer a means of communication between the penalty operator and the crossover and mutation operators. Figure 6.3 illustrates the template concept using the 4-queens problem.

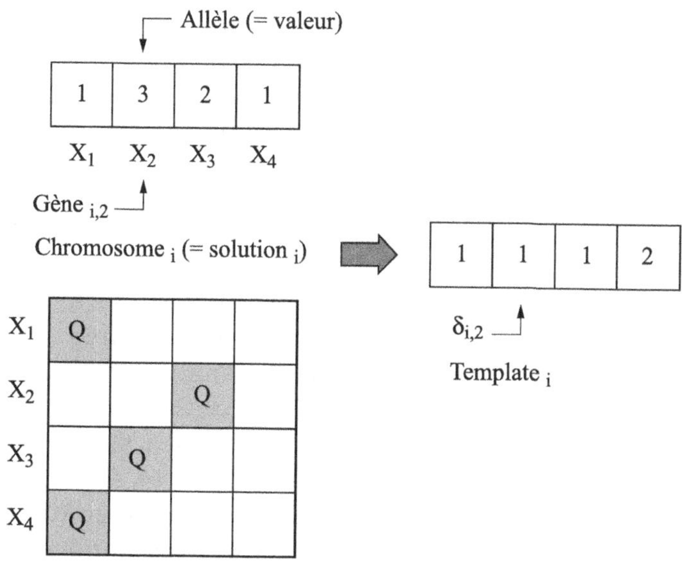

Figure 6.3. *Illustration of the templates using the 4-queens problem*

6.6.2. *The multi-agent model*

This model, developed by Ghédira and JLIfi [GHÉ 02], operates as follows: The initial population is divided into subpopulations so that each subpopulation contains the individuals that violate the same number of constraints. Then, each subpopulation will be assigned to an agent called a species agent. If n is the number of violated constraints relating to a species agent, then this agent is called $species_n$, n being the specificity.

So each species agent performs its own genetic algorithm guided by both the min-conflict heuristic and the templates. A so-called interface agent was added to this population of agents to act as an intermediary between this population and the user. It is responsible, on the one hand, for the creation of the species agents and, on the other hand, for detecting the best solution obtained by the interactions between the species agents.

Given a CSP, the interface agent randomly generates an initial population of chromosomes and does so in a manner similar to the traditional genetic algorithms. Subsequently, it divides it into subpopulations according to the number of constraints violated by the chromosomes. Finally, it creates a $species_n$ agent for each subpopulation of specificity n and requests it to start the execution of its genetic algorithm.

Before starting its optimization process, each $species_n$ agent sets all the templates (corresponding to its chromosomes) to zero using the penalization operator. Thereafter, it starts its genetic process on its initial subpopulation. This process then produces a subpopulation *pop*, which has only once been submitted to the crossover and mutation processes that correspond to a generation. For each *pop* chromosome, $species_n$ calculates the number of ncv constraints violated. If this number is equal to the specificity of the species to which the chromosome belongs, namely n, then this replaces one of its parents, which will be randomly selected. Otherwise, two types of processing are possible:

– If there is another species with this number ncv as a specificity, then this chromosome will be transmitted to this species.

– Otherwise, this chromosome will be sent to the interface agent that then creates a new species agent with ncv as a specificity and sends it the chromosome in question. Subsequently, the interface agent informs all the other species agents of this creation and requests that the new species, which has ncv as a specificity, starts its optimization process.

It is important to note that all agents give top priority to the processing of received messages. Indeed, if an agent receives a message, it stops its current behavior, saves the current context, processes the message, updates its local knowledge, and subsequently returns to its last context and behavior.

The species agents will carry on with their behaviors until the stop criterion specified by the user is achieved. This stop criterion, in fact, corresponds to the number of generations that each agent must not exceed. It can differ from one agent to another.

If, at the end of their behaviors, all the $species_n$ agents have not encountered any chromosome that violates zero constraints, then they transmit one of their chromosomes – randomly selected – to the interface. The interface agent then determines the best of these chromosomes (which violates the minimum of constraints) and provides the user with it.

If one of the species agents meets a chromosome that violates zero constraints, it transmits it to the interface agent that, in turn, informs all the other species agents of it so that they stop their genetic processes.

These interactions are detailed in algorithms 6.5–6.11.

6.6.3. *Genetic process*

This process is different from the canonical genetic algorithm (GA) in its use of templates and the min-conflict heuristic. It starts by determining the *mating-pool*, which includes pairs of chromosomes that are randomly selected using the *put-into-pairs* procedure. From each pair of chromosomes, the crossover operator produces a single child chromosome. The crossing over of two parent chromosomes will only take place if a random number smaller than the cross-over probability is generated. This probability is set by the user at the start of processing.

Using the new data structure "template", a generated child inherits the best genes, that is the "strongest" genes, from its parents. The probability for a parent chromosome $chromosome_p$ ($p = p_1$ or p_2) to propagate its $gene_{p,q}$ to its child chromosome is equal to $1-$ ($template_{p,q}$ / Sum), where $Sum = template_{p1,q} + template_{p2,q}$. This confirms the fact that the strongest genes, that is those violating the fewest constraints, have more chance than the others of being inherited by the child chromosome.

For each of its chromosomes selected according to the mutation probability P_{mut}, the $species_n$ uses the min-conflict heuristic. First, it is a matter of traversing the template corresponding to the chromosome and identifying the gene (variable) that is involved in the maximum number of violated constraints, then, selecting a new value for this gene from its domain, and finally instantiating this gene with this value.

It should be noted that if several genes with the same maximal weight value are detected, then one of these genes is randomly selected. If the chromosome has not been improved, insofar as the number of constraints violated remains the same, then this value is still kept so as to potentially explore new search areas. If the number of constraints violated by the chromosome in question increases, then the old chromosome is kept; otherwise, the new one is taken.

In cases where the chromosome obtained does not violate any constraint, $species_n$ requests that the interface agent stop the entire process. This then requests that all the species agents stop their behaviors and subsequently supply the user with the chromosome in question.

Should the need arise, if the new chromosome violates the same number of constraints, then it will be added to its subpopulation and $species_n$ will continue its behavior. Otherwise, that is, this chromosome violates a number of ncv constraints such that $ncv \neq n$, then $species_n$ checks if $species_{ncv}$ exists in its list of acquaintance agents. If this exists, it will send it the chromosome in question; otherwise, this chromosome will be sent to the interface agent that, in turn, will create a new $species_{ncv}$ agent and assign it to this chromosome.

1. m:= getMsg (mailBox)
2. case of m
   a. optimization process (sub population): apply behavior (sub population)
   b. take into account (chromosome): population-pool:= population-pool {chromosome}
   c. inform new agent ($species_{ncv}$): list acquaintances:= list acquaintances $\cup$ {$species_{ncv}$}
   d. stop process: stop behavior

Algorithm 6.5. *The processing of a message by a species agent*

The messages exchanged between the agents are the following:

– SendMsg (source, recipient, "message"): "message" is sent by "source" to the "recipient".

– ReceiveMsg (mailBox) processes the first message found in the mailBox.

Apply behavior (initial population)

```
init local knowledge
For i:= 1 to number of generations do
   update template (initial population)
   pop:= generic process (initial population)
   For each chromosome ∈ pop do
   ncv:= compute violated constraints (chromosome)
   If ncv = n then
       Replace by (chromosome)
   Else if Exist-ag (species_ncv) then
     sendMsg (species_n, species_ncv, ' take into account (chromosome)')
   Else
     sendMsg (species_n, Interface, ' create agent (chromosome)')
   End if
  End for
End for
sendMsg (species_n, Interface, ' result (one chromosome, specificity)')
```

Algorithm 6.6. *Behavior of a species$_n$ agent*

Genetic-process

```
mating pool;= matching(population pool)
update template (mating pool)
offspring pool crossed:= crossing (mating pool)
offspring pool mutated:= mutating (offspring pool crossed)
Return (offspring pool mutated)
```

Algorithm 6.7. *The genetic process*

Crossing (mating-pool)

```
If (size (mating-pool) < 2) then
  Return (mating-pool)
Else
  For each pair ∈ mating-pool do
    If (random [0, 1] < P_cross) then
      offspring = cross over (first pair, second pair)
      ncv:= compute violated constraints (offspring)
      If ncv = 0 then
        sendMsg (species_n, interface, ' stop process (offspring)')
      Else
        offspring pool:= offspring pool ∪ {offspring}
      End if
    End if
  End for
  Return (offspring pool)
End if
```

Algorithm 6.8. *The cross-over process*

Cross-over (chromosome$_{p1}$, chromosome$_{p2}$)

```
For q:= i to size (chromosome_p1) do
    sum:= template_p1,q + template_p2,q
    if (random integer [0, sum- 1] < template_p1,q) then
      gene_p3,q:= gene_p2,q
    Else
      gene_p3,q:= gene_p1,q
    End if
  End for
  Return (chromosome_p3)
```

Algorithm 6.9. *The cross-over operator producing the child p3*

It is worth noting that in the particular case of a monochromosomic population, that is population containing a single chromosome, the genetic process starts directly via the mutation operator. It should be noted that this state will not persist. Indeed, the size will increase due to the chromosomes transferred while generating other species.

The operation of the crossover operator is illustrated, with different iterations, in Figure 6.4 on the 4-queens example from section 1.2.2. If we cross the two chromosomes *chromosome*$_1$ and *chromosome*$_2$, we will need to use the templates *template*$_1$ and *template*$_2$ respectively. To generate gene$_{3,1}$ for the child chromosome, the sum of template$_{1,1}$ and template$_{2,1}$ has been calculated. If we generate zero as a random number between 0 and *Sum*-1, then since zero is less than template$_{1,1}$ (equal to 1) we take the value of the gene from chromosome$_2$ as indicated in Figure 6.4. Processing is continued until the last gene to generate the child chromosome *chromosome*$_3$. By calculating its number of violated constraints, we find that it violates a single constraint.

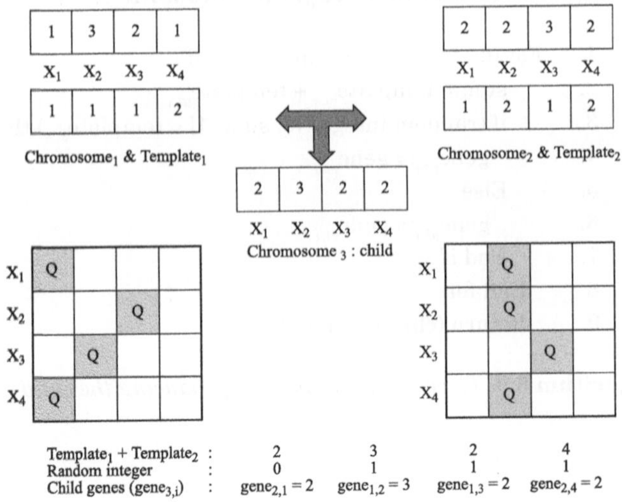

Figure 6.4. *Cross-over operator applied to the 4-queens example*

Mutating (offspring-pool)

```
For each chromosome ∈ offspring pool do
  If (random [0, 1] < P_mut) then
    gene_p,q := Min conflict heuristic (chromosome_p)
    ncv:= compute violated constraints (chromosome_p)
    If ncv = 0 then
      sendMsg (species_n , interface, ' stop process (chromosome_p)')
    Else
      offspring pool mutated:= offspring pool mutated ∪ {chromosome_p}
    End if
  End if
End for
Return (offspring pool mutated)
```

Algorithm 6.10. *The mutation process*

Min conflict heuristic (chromosome$_i$)

```
δ_ij := Max (template_i)
ncv*:= n /* n: the total number of constraints*/
For each v ∈ D_j do
  ncv:= compute violate constraints (v)
  If ncv < ncv* then
    ncv*:= ncv
    v*:= v
  End if
End for
value (gene_i,j):= v*
update (template_i)
Return (ncv*)
```

Algorithm 6.11. *The min-conflict function*

The operation of the mutation operator is illustrated in Figure 6.5 using the 4-queens example. We propose mutating chromosome$_2$ from Figure 6.4.

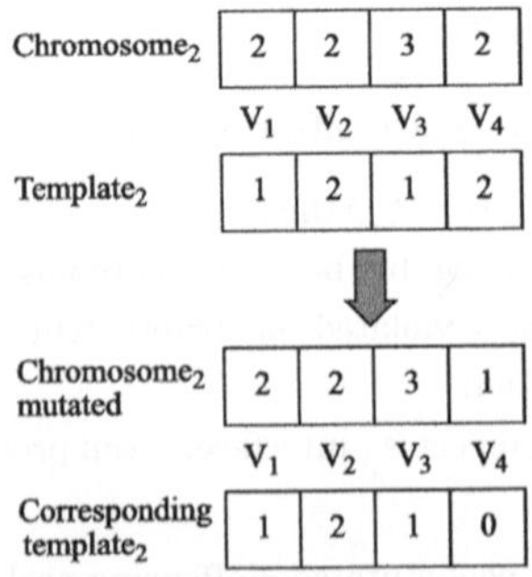

Figure 6.5. *Mutation operator applied to the 4-queens example*

Based on the template, the first step of this operator will give us the choice between $gene_{2,2}$ and $gene_{2,4}$ given that both have a maximum template value of 2. Let us assume that the guidance operator will guide us toward the fourth $gene_{2,4}$ to which the min-conflict heuristic will be applied. This heuristic will instantiate this gene with the value 1, given that it is the value that minimizes the number of constraints violated by the $chromosome_2$ (value equal to 0).

6.6.4. *Extensions*

6.6.4.1. *Distributed double-guided genetic algorithm*

The objective of the distributed double-guided genetic algorithm, denoted D2G2A, is to further diversify the search in relation to DG2A in order to increase the chances of escaping the trap of local optima. The simplest mechanism to diversify the search is, undoubtedly, the addition of noise during the search process.

In traditional genetic algorithms, it is the mutation that is used to create diversity in a population and thus to avoid degeneration of the population. In DG^2A, the mutation operator is improperly named since it is, in fact, an operator that improves the chromosome in question and does so by changing the allele (value) that violates the most constraints. This risks eliminating certain values that could, perhaps, orient us to the most promising search area. It is, therefore, necessary, from time to time, to cause mutations that can potentially generate some of these "potentially promising" values.

Hence, the idea to create a new parameter called *guidance probability* and hence the development of a new search dynamic.

In this dynamic, only the mutation process will be different from that of DG^2A and the other processes remain unchanged. That which is new lies in the mutation subprocess. Indeed, for each of the chromosomes selected according to the mutation probability P_{mut}, the guidance probability P_{guid} is also applied. The use of the min-conflict heuristic by the $species_n$ then becomes a random process of probability P_{guid}. This dynamic is illustrated in algorithms 6.12–6.14.

Mutating (offspring pool)

```
For each chromosome ∈ offspring pool do
  If (random [0, 1] < P_mut) then
    If (random [0, 1] < P_guid) then
        Guided Mutation (chromosome_p)
    Else
        Random Mutation (chromosome_p)
    End if
    ncv:= compute violated constraints (chromosome_p)
    If ncv = 0 then
      sendMsg (species_n, Interface, ' stop process (chromosome_p) ')
    Else
      offspring pool mutated:= offspring pool mutated ∪ {chromosome_p}
    End if
  End if
End for
Return (offspring pool mutated)
```

Algorithm 6.12. *New mutation process*

Guided Mutation (chromosome$_p$)

1. gene p,q:= Min conflict heuristic (chromosome$_p$)
2. Return (chromosome$_p$)

Algorithm 6.13. *The guided-mutate function*

Muter-aleatoire (chromosome$_p$)

1. Choose randomly a gene$_a$
2. Choose value randomly a value v ∈ domain (gene$_a$)
3. value (gene$_a$):= v
4. Return (chromosome$_p$)

Algorithm 6.14. *The random-mutate function*

6.6.4.2. *Dynamic distributed double-guided genetic algorithm*

The idea of this algorithm, denoted D^{3}G^{2}A, is to make the calculation of the genetic parameters dynamic so they evolve over time. The cross-over P_{cross} and mutation P_{mut} operators of each species agent will be calculated according to the number of constraints violated and a weighting coefficient ε strictly between 0 and 1.

Therefore, once the species$_n$ agent has been created, it will proceed to the calculation of its own cross-over and mutation operators before proceeding to optimization via its own genetic algorithm. Each species$_n$ agent will then have a new task. This task is to check the number *n* of constraints that it violates compared with the total number *N* of the problem's constraints. There are three possible cases:

1) *n* is high: in this case, the species$_n$ agent is responsible for a subpopulation that will be classified as weak. It is therefore advantageous for the agent to reduce the cross-over probability P_{cross} in order to decrease the chances of weak

individuals classified from crossing over. It is rather more advantageous for this agent to diversify the search to escape this bad area by favoring mutation with the hope of reaching a more promising area,that is an area containing values that violate fewer constraints. This is why it will increase its mutation probability P_{mut}. Then the operators will be:

$$P_{crosn} \leftarrow P_{cross} * \varepsilon$$

$$P_{mutn} \leftarrow P_{mut}/\varepsilon$$

2) n is low: in this case, the species$_n$ agent is assigned a subpopulation that will be classified as strong. It is therefore advantageous for it to increase its crossover probability P_{cross} in order to increase the chances of such individuals crossing over and potentially staying in the right search area. It is not advantageous for this agent to mutate its individuals, for fear of creating a mutation that violates more constraints. This is why it will reduce its mutation probability P_{mut}. Then, the operators will be:

$$P_{crosn} \leftarrow P_{cross}/\varepsilon$$

$$P_{mutn} \leftarrow P_{mut} * \varepsilon$$

3) n is average: in this case, the species$_n$ agent will keep the original parameters.

In what follows, n is considered to be high if $n > N / 2$, low if $n < N / 2$, and average if $n = N / 2$.

The following algorithm describes the new genetic process (line 2 shows the calculation of the operators according to the three preceding cases).

Generic process

1. mating-pool:= matching (population pool)
2. (P_{crosn}, P_{mutn}):= compute operator (P_{cross}, P_{mut})
3. offspring pool crossed:= crossing (mating pool)
4. update template (offspring pool crossed)
5. offspring pool mutated:= mutating (offspring pool crossed)
6. Return (offspring pool crossed)

Algorithm 6.15. *Genetic process relating to D^3G^2A*

6.7. Bibliography

[BOU 03] BOUAMAMA S., GHÉDIRA K., "D^2G^2A and D^3G^2A a new generation of distributed guided genetic algorithms for Max-CSP", *AITOCP Artificial Intelligence Techniques for Optimization and Constraint Problems a Joint Conference of the 7th World Multi-Conference on Systemics, Cybernetics and Informatics*, Orlando, Florida, FL, July 27–30, 2003.

[DAR 59] DARWIN C., *The origin of Species*, 6th ed., John Murray, London, 1859.

[FRE 92] FREUDER E.C., WALLACE R.J., "Partial constraint satisfaction", *Artificial Intelligence*, vol. 58, pp. 21–70, 1992.

[GHÉ 93] GHÉDIRA K., MASC: une approche Multi-Agents des Problèmes de Satisfaction de contraintes, Doctoral Thesis, ENSAE, Toulouse, France, 1993.

[GHÉ 94] GHÉDIRA K., "Distributed simulated reannealing for dynamic constraint satisfaction problems", *Proceedings of the International Conference on Tools with Artificial Intelligence*, New Orleans, LA, USA, 1994.

[GHÉ 96] GHÉDIRA K., "A distributed approach to partial constraint satisfaction problem", in PERRARRI J.W., MUILER J.P. (eds), *Lecture Notes in Artificial Intelligence, Distributed Software Agents and Applications*, vol. 1069, Springer Verlag, Berlin Heidelberg, 1996.

[GHÉ 02] GHÉDIRA K., JLIFI B., "A distributed guided genetic algorithm for max-csp", *Revue d'intelligence artificielle*, vol. 16, no. 3, pp. 1–16, 2002.

[GHÉ 07] GHÉDIRA K., *Optimisation combinatoire par métaheuristiques*, TECHNIP, July, 2007.

[GOL 94] GOLDBERG D.E., *Algorithmes génétiques: exploration, optimisation et apprentissage automatique*, Addison Wesley, France, 1994.

[HOL 75] HOLLAND J.H., *Adaptation in Natural and Artificial Systems*, University of Michigan Press, Ann Arbor, 1975.

[LAN 60] LAND A., DOIG A., "An automatic method of solving discrete programming problems", *Econometrica*, vol. 28, no. 3, pp. 497–520, 1960.

[LAU 98] LAU, T.L., TSANG, E.P.K., "Solving the radio link frequency assignment problem with the guided genetic algorithm", *Proceedings of NATO Symposium on Radio Length Frequency Assignment, Sharing and Conservation Systems (Aerospace)*, Aalborg, Denmark, October 1998.

[MIN 92] MINTON S., JOHNSTON M., PHILIPS A., LAIRD P., "Minimizing conflicts: a heuristic repair method for constraint satisfaction and scheduling problems", *Artificial Intelligence*, vol. 58, pp. 161–205, 1992.

[MIN 94] MINTON S., JOHNSON M.D., PHILIPS A.B., LAIRD P., "Minimizing conflicts: a heuristic method for constraint-satisfaction and scheduling problems", *Artificial Intelligence*, vol. 58, pp. 161–205, 1992. Reprinted in constraint-based reasoning, in FREUDER E.C., MACKWORTH A.K. (eds), MIT Press, MA, 1994.

[TSA 93] TSANG E., *Foundations of Constraint Satisfaction*, Academic Press, 1993.

[TSA 98] TSANG E.P.K., LAU T.L., *Solving the Radio Link Frequency Assignment Problem with the Guided Genetic Algorithm*, University of Essex, Colchester, UIC, 1998.

[WAL 96] WALLACE R.J. "Enhancement of branch and bound methods for the maximal constraint satisfaction problem", *Proceedings of the National Conference on Artificial Intelligence*, Portland, Oregon, 1996.

[YOK 94] YOKOO M., "Weak-commitment search for solving constraint satisfaction problems", *Proceedings of AAAI'94*, pp. 313–318, 1994.

Chapter 7

Constraint Satisfaction and Optimization Problems

In many real-world applications, there is a need to find the best possible solution to a problem and not only a solution that satisfies all of the constraints. This is what justifies the extension of the CSP framework to the CSOP framework via the introduction of a *cost* function whose value depends on the values assigned to the variables.

Compared to CSPs that regard all potential solutions as good solutions, the CSOPs explore all of these solutions, evaluate their respective costs in relation to an economic function called *cost* and only retain those offering optimal cost. In the case of max-CSP, the cost – also called objective function – consists of maximizing the number of satisfied constraints. The CSOPs are doubly complex to the difficulties related to complete satisfaction of constraints and are thus added to those relating to optimization. In what follows, we define the CSOP formalism as well as the different resolution techniques associated with it.

7.1. Formalism

DEFINITION 7.1:– CONSTRAINT SATISFACTION AND OPTIMIZATION PROBLEM: *Formally a CSOP is a quintuplet* (X, D, C, R, f) *defined by*:

– a set of n variables: $X = \{X_1, X_2, \ldots, X_n\}$;

– a set of n discreet and finite domains: $D = \{D_1, D_2, \ldots, D_n\}$ where D_i is the set of values associated with variable X_i;

– a set of m constraints: $C = \{C_1, C_2, \ldots, C_m\}$ where any constraint C_i concerns a subset of variables;

– a set of m relations $R = \{R_1, R_2, \ldots, R_m\}$ where each relation R_i is defined by a subset of the Cartesian product $D_{i1} \times D_{i2} \times \ldots \times D_{ik}$ corresponding to the set of possible value combinations for C_i, k being the number of variables involved in C_i;

– a *cost* or *objective* function defined by:

f: $S_p \rightarrow$ numerical value where S_p is the set of the potential solutions.

Thus, a *CSOP* (X, D, C, R, f) is a *CSP* (X, D, C, R) that seeks complete and consistent instantiations optimizing a *cost* function f.

7.2. Resolution methods

As it is still not known whether class P (all of the problems that can be solved in polynomial time) can be identified by class NP (all non-deterministic polynomial problems) and the CSOPs like the CSPs are NP complete, there is not (yet?) a general algorithm that could be used to solve a CSOP in polynomial time. This has motivated many researchers to make these both important and complex problems the focus of their research. Many resolution methods have thus been developed both in OR and AI. These methods may be broadly classified into two large categories: exact methods that ensure the completeness of the resolution but require much temporal complexity and approximate methods that sacrifice completeness to gain in temporal complexity:

– *Exact methods*: Despite the optimal quality of their results, these approaches generally encounter difficulties with applications of a significant size. In this category, we discuss the most traditional method: branch and bound. Note that the exact methods set out in the max-CSP framework can easily be adapted to the case of the CSOPs.

– *Inexact methods*: These methods are especially used by practitioners that deal with large-scale optimization problems where rapid resolution is required and optimality is not of primary importance. We present, within this framework, the *tunneling* algorithm. Note that the inexact methods set out in the max-CSP framework can be easily adapted to the case of the CSOPs. For more details, the reader is referred to [GHÉ 07].

7.2.1. *Branch-and-bound algorithm*

The version of the depth-first branch-and-bound algorithm is the natural extension of the backtracking algorithm within the CSOP framework. It is an algorithm that performs a depth search to supply one or several optimal solutions from the set of potential solutions.

Given that the CSOP is an NP-hard problem, the search for an optimal solution by the branch-and-bound algorithm is meticulous in terms of computation time. Hence, the idea of using a heuristic to guide this search and find an optimal solution in a reasonable computation time.

Similar to section 6.1, it is essentially a matter of pruning useless search areas depending on whether the current node can lead to a solution that has a more optimal cost than the best solution already found. To do this, the algorithm has a *bound* that corresponds to the value for the most optimal solution found to date. At the beginning of the resolution, this bound is set to $-\infty$ (case of maximization). After each node x is generated, the branch-and-bound algorithm decides if this node must be rejected depending on whether it leads to a better solution or not. For this, it calculates for each current node x the value obtained by the cost heuristic function f, which is the sum of:

– $g(x)$: the cost from the starting node (root) to the current node x;

– $h(x)$: the heuristic maximal estimate of the cost that could induce the variables not yet instantiated.

If f is less than the *bound*, no extension of the partial instantiation is feasible and the entire sub-branch is abandoned otherwise the new value (of the current variable) is accepted and the traversal of this sub-branch continues. In what follows, we outline the branch-and-bound algorithm.

Branch & Bound (X, D, C, R, f, h)
Begin

1. Bound:= - ∞
2. S*:= Ø
3. S*:= B N B (X, {}, D, C, f, h)

End

Algorithm 7.1. *The branch-and-bound algorithm*

Note that the efficiency of the branch-and-bound algorithm strongly depends on the choice of two factors: the heuristic h and the initial value of the *bound*. A wealth of studies has been developed in this framework and added to its literature.

One of the most common techniques within this framework is known by the term *iterative deepening*, which proceeds by successively testing a bound of maximal cost. With each iteration, an in-depth tree search is launched with the additional constraint that the solution must not exceed this bound. Pruning is accomplished in the same way as for branch and bound. If, by the end of the search, no solution has been found, the value of the maximal bound is reduced and a new search is attempted and this happens until a new solution is found.

Compared to branch and bound, this method cuts many more branches in each iteration but, each time, it runs the risk of missing the best solution and having to restart a search with a new bound, and thus generating nodes already explored during previous unsuccessful searches.

```
Procedure B N B (V, I, D, C, f, h)
Begin
    If V = Ø then
        If f(I) > Bound then
            Bound:= f(I)
            Return (I)
        End if
    Else
        If h(I) > Bound then
            Let x ∈ V
            For each v ∈ D_x do
                If I ∪ {(x, v)} consistent then
                    B N B (V\{x}, I ∪ {(x,v)}, D, C, f, h)
                End if
            End for
        End if
    End if
End
```

Algorithm 7.2. *The BNB procedure in the case of maximization*

7.2.2. *Tunneling algorithm*

Traditionally, the local search is an effective tool against problems considered to be very difficult. This success has prompted many researchers to develop variants of these methods, which include the tunneling algorithm. This algorithm, an extension of the GENET method, is due to [VOU 94].

The most important aspect of this algorithm is that it has introduced an iterative and dynamic modification mechanism of the *cost* function, which can be used to avoid the trap of local optima. This mechanism mainly consists of operating a tunneling function that reflects the characteristics of the objective function and controlling it by means of penalties.

As is commonly known, the objective function always depends on both the problem to be solved and its characteristics. In this respect and to reserve a general application framework for the tunneling algorithm, C. Voudouris and E. Tsang define a generic cost function, which incorporates the costs of the constraints and the costs of the instantiations. These parameters form the core of the *Tunneling* algorithm. They are defined as follows.

7.2.2.1. *Constraint costs*

In practice, the hard constraints that reflect obligations and must imperatively be satisfied must have a cost that reflects their importance compared to the soft constraints that represent preferences and are to be satisfied as best as possible. For this, if r_k is the cost of violation for the constraint C_k and S a complete instantiation, then:

$$C_k(S) = r_k \text{ if } C_k \text{ is violated by } S, 0 \text{ otherwise.}$$

This cost is called the primary cost of the constraint C_k. For the tunneling algorithm, it is increased by a penalty p_k:

$$C_k^T(S) = r_k + p_k \text{ if } C_k \text{ is violated by } S, 0 \text{ otherwise.}$$

At the beginning of the search, $\forall\ C_k$ we have $p_k = 0$. The increase of the penalties is achieved by the algorithm when the process becomes trapped in a local optimum.

7.2.2.2. *Instantiations costs*

The instantiation of a variable X_i by different values val_i from its domain D_i can infer different costs. Thus, a primary cost $a_{ij} \in N$ is associated with each instantiation $\{(X_i, \text{val}_j)\}$, which includes the cost resulting from the assignment of val_j to X_i:

$$L_{ij}(S) = a_{ij} \text{ if } \{(X_i, \text{val}_j)\} \in S, 0 \text{ otherwise}$$

For the tunneling version, this cost is increased by a penalty p_{ij}:

$$L_{ij}^T(S) = a_{ij} + p_{ij} \text{ if } \{(X_i, \text{val}_j)\} \in S, 0 \text{ otherwise}$$

7.2.2.3. *Objective function*

The objective function for the CSOPs may be defined as follows:

$$g\ (S) = \sum_{i=1\ldots|Z|} \sum_{j=1.|Dxi|} L_{ij}(S) + \sum_{k=1.|C|} C_k(S)$$

The tunneling version adds penalties to the constraints and instantiations:

$$g^T(S) = \sum_{i=1.|Z|} \sum_{j=1.|Dxi|} L_{ij}^T(S) + \sum_{k=1.|C|} C_k^T(S)$$

This tunneling $g^T(S)$ function is therefore dynamically modified by the tunneling algorithm so as to force the search to move away from a local optimum by means of a penalty mechanism, which will be detailed later in this section.

As already mentioned, the tunneling algorithm includes a local search method. This method, *GENERATE*, is based on the following reasoning. Starting with a randomly selected initial state and at each iteration it calculates the valuations g_v generated by assigning each value val_{kj} to a variable X_k. The instantiation $\{(X_k, \text{val}_{kj})\}$ that gives the best cost will be accepted and incorporated in the initial configuration. When the value of the objective function is the same for two consecutive iterations, it can be concluded that it is of a local optimum.

To escape this situation and have more chances of achieving a global optimum, the local optimum is penalized by an additional cost, modifying the objective function and thus enabling other alternatives to be explored.

This principle was implemented by two versions of the tunneling algorithm: one stage (ST) and two stage (2ST).

One-stage tunneling: The one-stage algorithm is based on the following procedure:

We can observe that GENERATE uses the g^T function instead of g. The g^T function provides a higher cost than g due to the penalty that is added to the primary cost, which increases the gap between the objective function and the tunneling function and thus skews the search from the first encountered local optimum. To overcome this disadvantage, there is two-stage tunneling that consists of periodically replacing the tunneling function with the objective function.

```
Procedure One-Stage (X, D, g, g^T, S*)
Begin
    k:= 0
    S_k:= aleatory solution
    cost*:= g (S_k)
    S*:= S_k
    Repeat
        Repeat
            Minloc:= False
            GENERATE (X, D, g, S_k, S_{k+1})
            If cost* > g (S_k) then
                cost*:= g (S_{k+1})
                S* = S_{k+1}
            End if
            If g (S_{k+1}) = g (S_k) then
                Minloc = True
            End if
            k:= k + 1
        until (Minloc)
        Increment penalties (S_k)
    until (stopping condition)
    Sol-opt:= S*
End
```

Algorithm 7.3. *The one-stage procedure*

Two-stage tunneling: The two-stage version is similar to the one stage. The only difference is that 2ST includes two periodic steps: a step for minimizing the objective function and a tunneling step that optimizes the tunneling function.

The algorithm begins with a minimization step. Once it achieves a local optimum, it increases the penalties and changes its state.

```
Procedure GENERATE (X, D, state_i, state_i+1)
Begin
    state:= state_i
    set*:= Ø
    For each x_i ∈ X do
        state:= state - {(x_i, val_ij)}
        For each val_ij ∈ D_i do
            g_v:= g (state + {( x_i, val_ij)}
        End for
        set*:= val_ij | g_v optimal
        val_ij:= aleatory value in set*
        state:= state + {(x_i, val_ij)}
    End for
    state_i+1:= state
End
```

Algorithm 7.4. *The GENERATE procedure*

The two-stage algorithm is based on the following procedure:

```
Procedure Two-stage (X, D, g, g^T, S*)
Begin
    k:= 0
    S_k:= aleatory solution
    cost*:= g (S_k)
    S*:= S_k
    stage:= minimization
    Repeat
        Repeat
            Minloc:= False
            If stage = minimization then
                GENERATE (X, D, g, S_k, S_{k+1})
            Else if stage = tunneling then
                GENERATE (X, D, g^T, S_k, S_{k+1})
            End if
            If cost* > g (S_k) then
                stage = minimization
                cost*:= g (S_{k+1})
                S*:= S_{k+1}
            End if
            If g (S_{k+1}) = g (S_k) then
                Minloc:= True
            End if
            k:=k + 1
        until (Minloc)
        Increment penalties
        If stage = minimization then
            stage = tunneling
        Else
            stage:= minimization
        End if
    until (stopping condition)
    Sol-opt:= S*
End
```

Algorithm 7.5. *The two-stage procedure*

Both algorithms rely on the *increment-penalties* (S) procedure, which reflects the following penalization mechanism:

Initially, all of the penalties are equal to zero and the tunneling function is equivalent to the objective function. After each local optimum, the *increment-penalties* (S) procedure is used to increase certain penalties. To do this, this procedure requires two key pieces of information: the primary cost of a term (cost of a constraint or instantiation) and the number of times it has already been penalized. The FCR_i that stands for *frequency-cost ratio* is associated with each term *i*.

Let min-FCR be the set of the terms representing the violated constraints and the instantiations that have the minimum FCR relating to a local minimum. From these terms, a subset *penalizeSet* is extracted. It contains the elements whose primary costs are the highest. After determining this set, increment penalties (S) modifies the penalties associated with each of these elements and updates their frequencies [VOU 94].

Increment penalties conform to the following procedure:

```
Procedure increment penalties (S_k)
Begin
  1.  T instantiations:= set of instantiations of S_k
  2.  T constraints:= set of violated constraints of S_k
  3.  T:= T instantiations ∪ T constraints
  4.  MinFCR:= set of elements of T having minimal FCR
  5.  penalizedSet:= the elements of MinFCR having minimal
      costs
  6.  For each element m in penalizedSet do
  7.      penalty m:= penalty m + P'_ij
  8.      frequency m:= frequency m + 1
  9.  End for
End
```

Algorithm 7.6. *The increment-penalties procedure*

Despite its incompleteness, this algorithm has been successfully applied to the problems of the traveling salesman and frequency assignment in mobile radio networks and has obtained very good results according to [VOU 94].

This is what prompted the authors to extend their work to a wide range of combinatorial optimization problems [TSA 98, TSA 99, VOU 98, VOU 96]. They thus proposed a generic metaheuristic: guided local search (GLS).

Just like its predecessor, GLS combines the descent method with a specific mechanism to modify the cost function dynamically. However, its innovative idea consists of defining the characteristics of each candidate solution and associating them with costs and penalties.

Indeed, given the optimization problem (E, g) defined by the set E of the configurations and the cost function g, the objective function that will be optimized by GLS (E, g) is expressed as follows:

$$h\,(s) = g(s) + \lambda * \sum p_i * L_i(S)$$

where:

– S is the candidate solution;

– λ is the parameter of the GLS algorithm;

– p_i is the penalty of a characteristic i, initially nil;

– L_i: is the indicator such that $L_i\,(S) = 1$ if S possesses the characteristic i, 0 otherwise.

After visiting a local optimum, the algorithm increases the penalties of certain characteristics, which modify the objective function and direct the search toward new candidate solutions. This penalization mechanism is based on key information, the usefulness of penalization called

Util_i. The usefulness of penalization is associated with each characteristic i of a local optimum S^*:

$$\text{Util}_i(S^*) = L_i(S^*) * C_i/(1 + p_i)$$

where:

– C_i is the cost of the characteristic i;

– p_i is the current penalty of the characteristic i.

Therefore, at the level of a local optimum, these are the characteristics whose usefulness of penalization is the highest, which will be penalized: $p_i := p_i + 1$.

The guided local search has led to several extensions, including the guided genetic algorithms [TSA 99, VOU 98], which emphasize the standard genetic algorithms by modifying the objective function with a penalization mechanism so as to try to escape the trap of local optima.

7.3. Bibliography

[GHÉ 07] GHÉDIRA K., *Optimisation combinatoire par métaheuristiques*, Editions TECHNIP, 2007.

[HAO 98] HAO J.K., GALINIER P., HABIB M., "Métaheuristiques pour l'Optimisation Combinatoire et l'Affectation sous Contraintes", *Version révisée pour la revue d'Intelligence Artificielle*, 1998.

[TSA 98] TSANG E., VOUDOURIS C., "Constraint satisfaction in discrete optimization", *Proceedings of Unicom Seminar on Constraint Satisfaction and Discrete Optimization Overview*, London, April 1998.

[TSA 99] TSANG E., WANG C.J., DAVENPORT A., VOUDOURIS C., LAU T.L., "A family of stochastic methods for constraint satisfaction and optimization", *Proceedings of the International Conference on the Practical Application of Constraint Technologies and Logic Programming*, London, pp. 359–383, 1999.

[VOU 94] VOUDOURIS C., TSANG E., Tunneling algorithm for partial CSP and combinatorial optimization problems, Technical Report CSM-213, Department of Computer Science, University of Essex, 1994.

[VOU 96] VOUDOURIS C., TSANG E., "Partial constraint satisfaction problems and guided local search", *Proceedings of Practical Application of Constraint Technology*, pp. 337–356, London, UK, April 1996.

[VOU 98] VOUDOURIS C., "Guided local search joins the elite in discrete optimization", *Proceedings of DIMACS* workshop *on Constraint Programming and Large Scale Discrete Optimization*, Rutgers, New Jersey, September 1998.

Chapter 8

Distributed Constraint Satisfaction Problems

Globalization of the economy and democratization of the Internet, boosted by the huge growth in information and communication technologies, have largely contributed to the expansion of numerous distributed architectures. This distribution is due to various reasons: physical, functional, etc. These architectures may correspond to diverse and varied activities: administrative, economic, industrial and others such as social networks and Web services. An example of a geographically distributed architecture is given in Figure 8.1. Other applications, particularly relating to distributed artificial intelligence, have also been reinforced and strengthened with the boom of computer networks and telecommunications. Hence:

– The need for a distributed management and decision support system suitable for these interdependencies and these complex relationships of competition and/or cooperation and/or coordination.

– The idea is to draw inspiration, on the one hand, from the CSPs to express these relations as constraints and on the other hand, by the multi-agent systems to take into account the distributed aspect and the autonomy and locality properties of the entities that form these architectures.

This is what has given rise to distributed CSP (DisCSP), the subject of this chapter. We begin by defining the DisCSP framework and then discuss the associated distributed consistency reinforcement and resolution techniques. It should be noted that in addition to the difficulties related to CSPs, DisCSPs induce new ones inherent in simulation and implementation such as asynchronous behaviors, cycles and communication problems.

As with CSP, several extensions have been proposed to enrich the DisCSP framework in terms of flexibility and genericity. The best-known references are the following: [YOK 93], [HIR 97], [HIR 00], [MOD 02], [BOW 06], [KAT 06], [PEA 07], [GRE 07], [FAL 08], [SIL 08b], [SIL 08a] and [BRI 09].

The most-known extensions are the following: distributed constraint satisfaction and optimization problems (DCSOP), distributed maximal constraint satisfaction problems (DMCSP), distributed partial constraint satisfaction problem (DPCSP) and more generally distributed constraint optimization problems (DCOP).

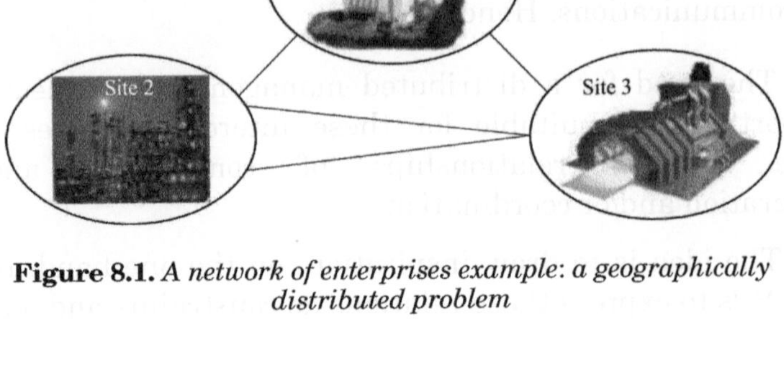

Figure 8.1. *A network of enterprises example: a geographically distributed problem*

8.1. DisCSP framework

This section discusses the DisCSP formalism and its extensions as well as the properties induced by the distributed aspect, namely distribution modes, communication protocols and convergence.

8.1.1. *Formalism*

DisCSP, represented in Figure 8.2, is a quintuplet (X, D, C, R, A) where:

– X is a finite set of n variables $\{X_1, X_2, \ldots, X_n\}$;

– D is the set of associated n domains $\{D_1, D_2, \ldots, D_n\}$;

– C is a finite set of m constraints each involving a subset of variables of X;

– R is the set of m relations associated with the m constraints;

– A is a finite set of p entities called agents $\{A_1, \ldots, A_p\}$.

In the majority of studies, each agent is regarded as an entity which is responsible for a non-empty subset of variables and their domains as well as all the constraints that connect them. Figure 8.2 formalizes the example from Figure 8.1 as a DisCSP where the sites are represented by agents, the data within the sites by variables linked via intra-agent constraints also called internal constraints and the relationship between sites by inter-agent constraints also called external constraints. We thus define a membership relation $\varphi: X \rightarrow A$ that associates each variable with an agent. Thus, each variable belongs to only one agent but an agent may be responsible for several variables.

This distribution of variables divides C into two separate subsets: the intra-agent constraints (C_{intra}) and the inter-agent constraints (C_{inter}).

The intra-agent constraints involve variables of the same agent: $C_{intra} = \{C_i \in C \mid \forall x_j, X_k \in C_i, \varphi(x_j) = \varphi(X_k)\}$.

The inter-agent constraints involve variables of different agents: $C_{inter} = \{C_i \in C \mid \exists x_j, X_k \in C_i, \varphi(x_j) \neq \varphi(X_k)\}$.

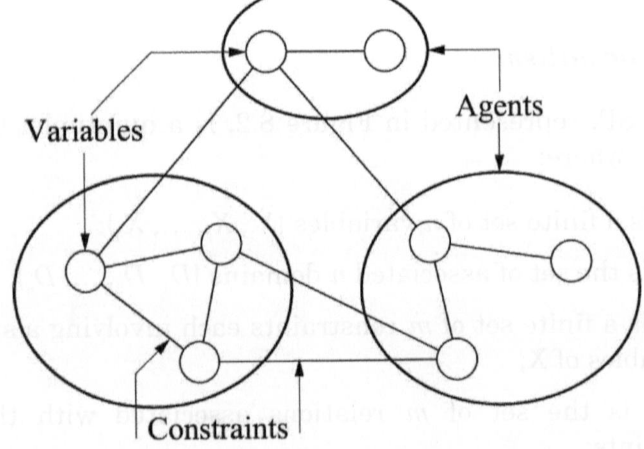

Figure 8.2. *Distributed Constraint Satisfaction Problem*

A solution to this problem is then an instantiation of all the variables, which satisfy both the intra-agent and inter-agent constraints.

Generally, an intra-agent constraint is only known by the agent that is responsible for the variables involved in this constraint and an inter-agent constraint is known by any agent that possesses at least one of its variables involved in this constraint. But we will see that this depends on the mode of distribution.

A DisCSP problem can also be seen as a set of p agents A_i where a sub-problem CSP P_i is assigned to each agent A_i and where the global solution is a "concatenation" of the local solutions to the p CSP.

Let us define the acquaintances of an agent A as the agents with which A shares at least an external (inter-agent) constraint.

8.1.2. *Distribution modes*

This section discusses the different modes of distribution for the problems under constraints. The interest in some of these distributions is comprehendible only by considering the associated problem resolution methods. Indeed, although equivalent, each of them possesses different properties that can have ample influence on these resolution methods.

8.1.2.1. *Implicit distribution*

This distribution, shown in Figures 8.2 and 8.3, is implicit in the sense that it strictly holds to both the formalism and the associated membership function φ described in section 8.1. Indeed there is no change at the level of the agents, namely their associated variables, domains, internal and external constraints. In Figure 8.3, the constraints C_{13}, C_{16}, C_{47} and C_{57} represent inter-agent constraints while the others are intra-agent constraints. This distribution is well suited to the naturally distributed problems insofar as it holds true to real life. Indeed, the agents of the DisCSP correspond to various real entities (sites shown in Figure 8.1) or virtual ones that correspond to the initial problem. Implicit distribution is the only one that does not require the problem to be redistributed.

8.1.2.2. *Canonical distribution with the addition of an agent*

Canonical distribution of a DisCSP, as shown in Figure 8.4, adds a new agent to the implicit representation. This agent is central to the system and possesses a copy of each of the inter-agent constraints. Thus, the resolution of this agent's sub-problem ensures that all inter-agent constraints are satisfied. Each agent is therefore connected to the new agent via an equality constraint linking each of its duplicated variables to its copy.

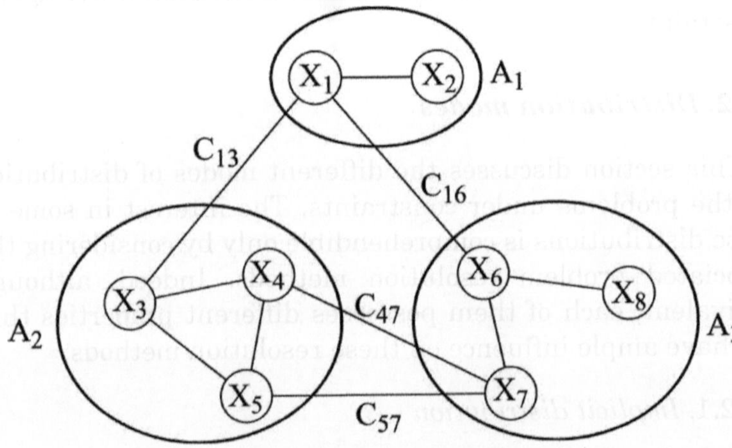

Figure 8.3. *The Implicit distribution mode*

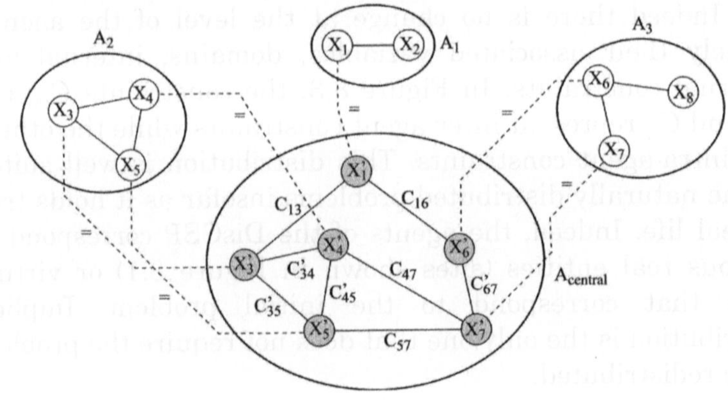

Figure 8.4. *The canonical distribution mode*

8.1.2.3. *Canonical distribution without the addition of an agent*

In Figure 8.5, the inter-agent constraints are "moved" inside each agent. This mode of distribution is also described as asymmetrical. Duplication is used for each of the external variables, that is those involved in inter-agent constraints. As discussed before, the duplicated variables are linked with their initial variables via equality constraints.

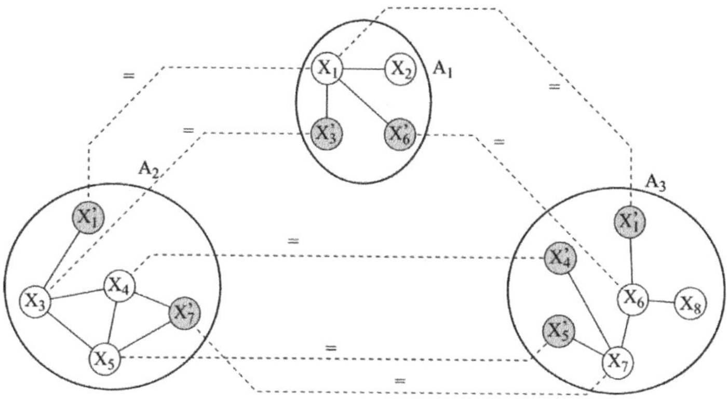

Figure 8.5. *The canonical mode of distribution without the addition of an agent*

8.1.2.4. *DOC distribution*

Ben Salah and Ghédira [BEN 02] proposed a generic framework, called DOC for distributed optimization under constraints, based on a distribution different from previous ones insofar as it fully complies with the basic principles of multi-agent systems, namely, locality, autonomy, asynchrony and behaviors performing in parallel. It is based on two types of agents: decision-maker agents and advisor agents.

– The decision-maker agents correspond to the initial agents of the problem, namely the agents defined in section 8.1. They are responsible for the instantiation of their internal variables while satisfying their internal constraints.

For the example in Figure 8.1, one decision-maker agent is associated per site.

– However, instead of a single central agent as is the case with canonical distribution, multiple agents called *advisors* coexist. Each advisor agent is in charge of satisfying the inter-agent constraints for which it is responsible. Furthermore, no intra-agent constraint is moved to the advisor agents as is the case in canonical distribution. This, on the one hand, allows both parallel resolution to be emphasized and resolution time to be reduced and on the other hand, allows centralization to be avoided and thus the multi-agent philosophy to be kept.

A global solution is obtained via interactions between the decision-makers and the advisors (see section 8.3.4). Creation of the decision-makers is immediate: for example, a decision-maker is associated with each subproblem (sub-CSP) or site. Determining the advisors is more crucial to the extent that it must distribute the search guidance to the fullest (in the form of advice provided by advisors to their acquaintance decision-makers) while minimizing inter-agent communication and ensuring global consistency.

We could imagine, for example, one advisor per external constraint but such a configuration would require centralized control mechanisms or very expensive negotiations and/or large-scale communications to ensure global consistency at the level of the boundaries, that is the variables involved in the external constraints.

Indeed, for the example in Figure 8.6, the configuration – an advisor for the constraint involving the two variables X_1 and X_2 and another advisor for the constraint linking X_1 to X_3 – would require a lot of negotiation to ensure the consistency of the value for variable X_1. For the example in Figure 8.7, the choice of a single advisor per pair of agents allows for easier detection of the nogood search directions, that is directions that do not lead to a solution without any unnecessary communication.

Taking these remarks into account, three basic configurations were proposed:

– One advisor per connected component (see Figure 8.6), a connected component being formed by nodes (variables) on the boundaries and the corresponding external arcs (external constraints).

– A single advisor per pair of agents (see Figure 8.7).

– In case of an isolated decision-maker, that is not connected to other decision-makers and consequently having no boundary, this decision-maker becomes its own advisor. It randomly selects a constraint from its internal constraints and regards it as a *virtual* external constraint (see section 8.3.5).

Depending on the nature of the boundary, a decision-maker may then have one or several advisors. Figure 8.8 shows an example of a decision-maker with two advisors.

Generally, any graph of agents formed by decision-makers and advisors can be represented as a linear combination of the basic subgraphs from Figures 8.6 and 8.7 and from the isolated decision-maker case.

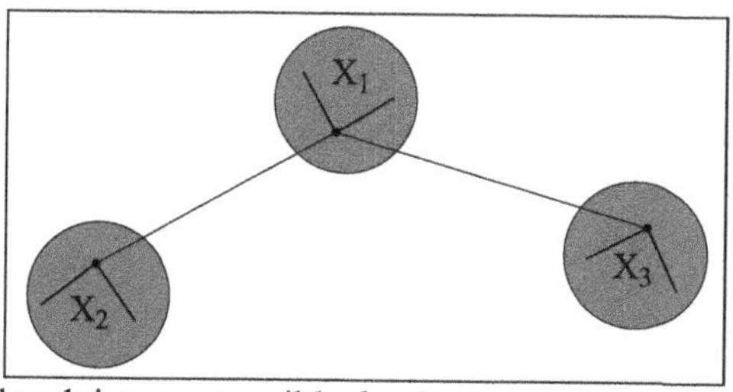

A single advisor responsible for the connected component (X_1, X_2, X_3)

Figure 8.6. *Example of a component connected on the boundaries*

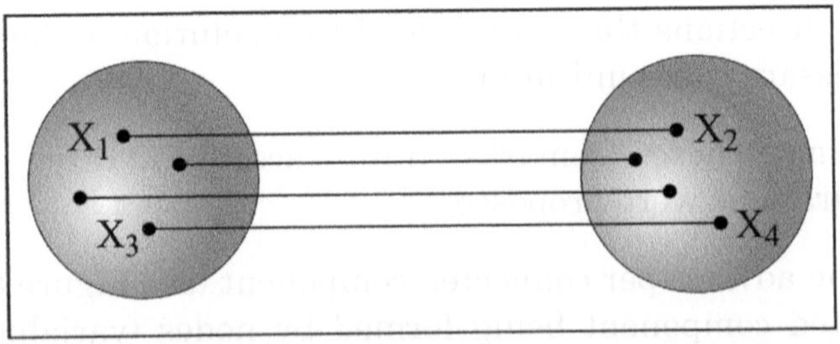

A single advisor responsible of all constraints between the two agents

Figure 8.7. *Example of an advisor responsible for several constraints linking two decision-makers*

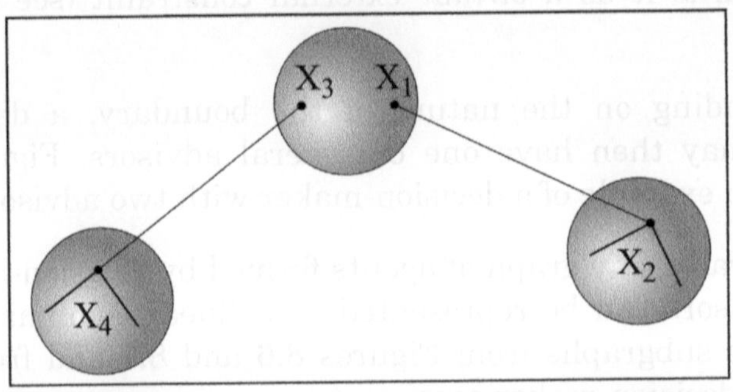

An advisor responsible for the constraint involving X_1 and X_2 and another advisor for the contraint involving X_3 and X_4

Figure 8.8. *A generic example of decision-makers and advisors*

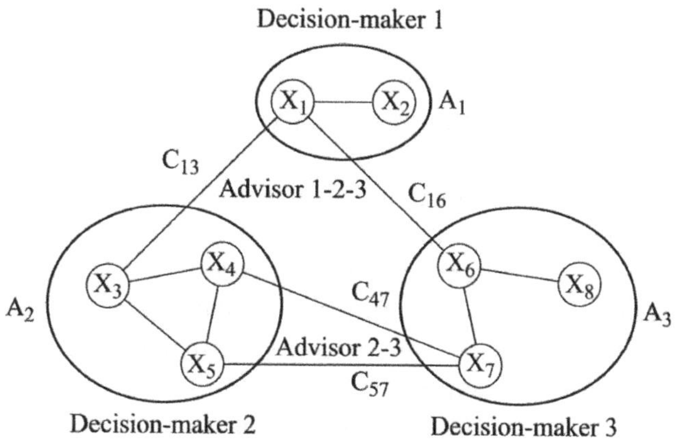

Figure 8.9. *The advisors from the example in Figure 8.3*

It is important to note that an external constraint may only be attached to one single advisor. On the other hand, an advisor can be responsible for several external constraints at once, even if they involve more than two agents. The DOC distribution from the example in Figure 8.3 is represented by Figure 8.9 where advisor1-2-3 advises the three decision-makers about the variables involved in constraints C_{13} and C_{16} while advisor2-3 advises decision-maker1 and decision-maker2 about the variables involved in constraints C_{47} and C_{57}.

8.1.3. *Communication models*

Within its literature, there are two types of communication models: the shared memory model and the message passing model. Each model has its advantages and disadvantages.

8.1.3.1. *Shared memory model*

The shared memory model, often referred to as parallel random access memory (PRAM), combines various processing units (processors or agents) via a shared memory. If the processors are distributed physically or geographically, then it is called distributed shared memory. One of the fundamental characteristics of shared memory is that all communication is implicit via the writing and reading of a global address space. A second characteristic is that synchronization is separate from communication: special synchronization mechanisms must be added to detect when data is produced or consumed.

Although this model allows for controlled parallelism and frees the programmers from the burden of an explicit location for shared data and from orchestrating inter-processor communication, it has some disadvantages mainly relating to the access-conflict problems. In practice, it is difficult to construct machines that provide uniform access (in terms of performance) to the memory beyond a certain number of processing units.

There are four types derived from the shared memory model PRAM, which differ in the particular management of the reading/writing operations:

– CRCW (concurrent read concurrent write): authorizes concurrent reading and writing to the same memory word;

– CREW (concurrent read exclusive write): authorizes concurrent reading but excludes simultaneous writing;

– ERCW (exclusive read concurrent write): authorizes concurrent writing but excludes simultaneous reading;

– EREW (exclusive read exclusive write): a single processor at a time is authorized to read or write in the same memory word.

8.1.3.2. *Message passing model*

The message passing model combines local memories and processing units. Inter-processor communication is carried out via a common medium (network, etc.) ensuring the transmission of messages.

Message passing is an explicit communication model. All communication involves explicit messages going from processors that have the data to those who require it. Furthermore, the arrival of a message at the destination processor constitutes an asynchronous event since certain programs must be invoked to deal with the messages. As a result, synchronization and communication are unified with the passing of messages.

The message passing model is a simple, intuitive, powerful and universal model since it provides a perfect adaptation of the simple local network of machines with supercomputers (multiprocessors).

8.1.4. *Convergence properties*

Self-stabilization is a convergence property of distributed algorithms: it allows the algorithms to be robust against dynamic environments and/or fault tolerant.

Consider a system with n nodes (processors, processes or agents) arranged according to a network architecture. A configuration expresses the global state, which is a concatenation of the local states for all of the nodes and contents of all the communication channels. The configuration space is divided into legal and illegal configurations. The legality of a configuration depends on the system's objective. In the case of DisCSP, a legal configuration corresponds to a solution to the problem. A protocol is defined as the set of transition functions for all of the nodes. A protocol (or distributed algorithm) is called self-stabilizing, if from any system configuration, it achieves, in finite time, a legal configuration in which to remain.

In [COL 99], the authors present a self-stabilizing and almost uniform asynchronous protocol for the resolution of the binary DisCSP, with one variable per node. The problem is defined by the authors as a network consistency problem. The underlying protocol (network consistency protocol) is based on a distributed version of graph-based backjumping and implemented on a spanning depth first search (DFS) tree. It uses a connectionist architecture where the communication is based on the shared memory PRAM-CREW. Logically, it is composed of two interleaved self-stabilizing sub-protocols:

– A sub-protocol for generating a spanning DFS tree;

– A sub-protocol for assigning values.

The second sub-protocol uses a privilege passing mechanism where a node is only activated if it is favored. Each node attempts to assign its variable a value consistent with those of its ancestors in the DFS tree and passes the privilege to its children. In case of failure, it passes the privilege to its parent. The complexity of the protocol is exponential with the DFS tree's height.

FKI and Ghédira [FKI 02] have proposed an improvement of the convergence speed of the second sub-protocol toward a solution to the problem. Acceleration is achieved by integrating filtering of the future variable domains, which allows for premature detection of inconsistency.Furthermore, a mechanism is added to accelerate the privilege's return to the node that is the source of conflict.

8.2. Distributed consistency reinforcement

Given the local nature of consistency properties, the consistency reinforcement algorithms are generally perceived as easy to parallelize/distribute. Although distributed algorithms seem to be similar to parallel processing methods for handling DisCSPs, research motivations are fundamentally different. The primary concern in parallelization is the efficiency, and we can choose any type of parallel computer architecture. In contrast, in distribution, there already exists a situation where knowledge about the problem (i.e. variables and constraints) is distributed among agents [YOK 95]. Therefore, the main research issue is how to reach a solution and how to reinforce consistency from this given situation.

The research effort has generally focused on distribution even if some pure parallelization studies exist such as:

– Samal and Henderson [SAM 87] gave parallel versions of AC-1, AC-3 and AC-4 and showed that any parallel algorithm for enforcing arc consistency (AC) in the worst case must have $O(nd)$ sequential steps, where n is the number of variables and d is the average size of domains. A computational model based on memory sharing (PRAM) and a multiprocessor machine was used to analyze these algorithms.

– Swain and Cooper [SWA 88] and Cooper and Swain [COO 88] discussed how to improve AC algorithms by exploiting parallelism and domain-specific problem characteristics. Thus, a massively parallel algorithm for AC was given and expressed as a digital circuit. The domain description is formalized as meta-predicates whose purpose is to represent invariant properties of domain problem instances. The meta-predicates are the input to an algorithm that produces an optimized AC chip that may have space complexity orders of many magnitude lower than the general-purpose AC chip. Besides, because of the close relationship between the sequential AC algorithm AC-4 and the AC chip, this domain-specific optimization technique can be used to reduce the time complexity of AC-4, by modifying AC-4 so that it simulates the optimized chip.

We focus now on some well-known algorithms that were developed to extend and adapt some centralized filtering algorithms for the distributed case. It should be noted that almost all the studies only deal with AC.

8.2.1. *The DisAC-4 algorithm*

Nguyen and Deville [NGU 95] presented DisAC-4, a coarse-grained parallel algorithm designed on the basis of AC-4 for a distributed memory computer using message passing communication. A method to efficiently exchange information between the cooperating processes was proposed. It uses k processes with $1 < k < n$, each one is called a "Worker"; n is the number of domains of the CSP. These processes run the same code but on different data.

Each *Worker* process handles a set of variables and associated domains. Like AC-4, computation of each Worker consists of two parts. In the first stage, it independently builds the local data structures *counter* and *Support*, and detects local inconsistent labels. In the second stage, the inconsistent labels are treated. They may be either locally generated inside the process or produced by the other workers. Hence, each Worker has to transmit its locally detected inconsistent labels to the other workers and collect inconsistent labels sent by them.

The space complexity of DisAC-4 is $O(n^2d^2)$. The List variable of AC-4 can be seen as a shared memory data structure, simulated by message passing. The other data structures of AC-4 are partitioned among the p workers.

In the first phase, the *Worker* needs all the domains but it only updates the domains of its variables. Note that the authors have adopted the hypothesis of fully connected processors architecture in which broadcast operation is available. Each process can send a message via any transmission channel. However, it can only receive a message via the channel that has been dedicated to it.

A *Controller* process is used, on the one hand, to detect the termination of the computation and on the other hand, to inform the other processes of this via a Stop message. The system stops when all the *Workers* have dealt with all their inconsistent labels and when no more inconsistent labels are generated.

8.2.2. *The DisAC-6 algorithm*

DisAC-6 is based on the DisCSP architecture defined in section 8.1.1. It was proposed by [HAM 02a]. It is an adaptation of the centralized algorithm AC-6 within the distributed framework. The key idea of this distributed algorithm consists in carrying out, on the one hand, local processing relative to the internal constraints of each agent and on the other hand, the processing concerning the explicitly shared constraints. But to process explicit constraints, a non-locality information problem arises. To confront this problem, the author suggests making the domains relating to the inter-agent constraints shareable via the sending of messages. For example, when filtering a domain D_i relating to constraint C_{ij}, the agent must consider the domain D_j that it does not have.

The DisAC-6 algorithm consists of two overall steps. The first is an initialization step for filtering the domains of each local variable and broadcasting the potential label removals to the concerned acquaintances. The second step corresponds to a phase of interactions where the local labels are questioned according to new removals.

Each agent begins by initializing its data structures. It tests viability for each of the variables that it possesses. Each arc (i, j) from the graph of constraints is tested with respect to the constraint C_{ij}. This processing is solely limited to the arcs such that X_i is a local variable. For this, the agent begins by seeking a viable (consistent) support for each value a for each of its variables X_i. If such a support does not exist, a must be removed from the domain of X_i, and then propagated respectively at a local and a non-local level.

After this initialization phase, the interaction phase can then begin. Each agent must then process the messages containing the labels removed by the other entities that it received. It determines the impact of these labels on its own data by propagating their removals at a local and a non-local level. For each label (X_i, a) detected as being non-viable, the agent must find a new support for the set of values supported by (X_i, a). This search is only carried out for the values of local variables. Thus, new removals can occur and subsequently be processed and propagated to the agents in question (acquaintances).

The worst case spatial complexity is around $O(n^2d)$. It is identical to that of the AC-6 algorithm. Regarding the temporal complexity, it is around $O(n^2d^3)$ with $O(nd)$ messages.

8.2.3. *The DisAC-9 algorithm*

This algorithm was also presented in [HAM 02a] to minimize the exchanged messages in the DisAC-6 algorithm. Each agent begins by determining the locally inconsistent labels. For each label (X_i, a) with X_i a local variable, the agent must decide whether to immediately inform its acquaintances about the removal of (X_i, a). For this, it explores the domain of each non-local variable X_j using its local knowledge on its domains, to estimate the impact of removing (X_i, a) on its domains. If (X_i, a) is the only support for (X_j, b), then the agent that possesses X_j must be informed of this removal. In the opposite case, the information is useless at this time and it can be delayed. In other words, when (X_j, b) has lost its only support (X_i, a), the agent possessing X_j is then informed not only about the removal of (X_i, a) but also about all removals carried out in the domain of X_i.

Using the constraint, bidirectionality property allowed the agents to partially share their information on the acquaintances: while $C_{ij}(a, b) = C_{ji}(b, a)$, any agent that possesses the variable X_i and not X_j also knows the constraint C_{ij}. This information can be used to infer some possible removals that can be carried out by the acquaintances. As a result, each message sent to an agent A of the distributed system causes for A the removal of at least one new value.

The spatial complexity of DisAC-9 is around $O(n^2d)$, while its temporal complexity is around $O(n^2d^3)$ with exactly nd messages point-to-point.

8.2.4. *The DRAC algorithm*

Proposed by [BEN 02a], this algorithm addresses both binary and n-ary problems.

The DRAC architecture is different from the previous algorithms. It associates one agent per constraint. It aims to simplify a CSP namely to reduce the variable domains via interactions between the constraint agents involving these variables.

Indeed, each constraint agent first determines the locally viable values for each of its variables (two in the binary case) and any value that does not allow any support will be removed.

This removal will lead to the appearance of new non-viable values and consequently new removals. The reduction of a domain relative to an agent leads to the (potential) reduction of one or several domains at the level of other agents.

The system reaches its stable state, when there are no longer any interactions between the agents and when no reduction is possible.

After the reduction, if the domain of a variable becomes empty, the algorithm detects the failure (absence of solution). Indeed, a message will be sent by the constraint agent that detected the failure to the interface agent to stop the process. Note that the interface agent acts as an intermediary between the constraint agents and the user.

For binary problems, the spatial complexity of this algorithm is the same as for AC-6. Indeed, each agent keeps a list of supports that, in the worst case, is of a size equal to $2d - 1$. Knowing that we have m agents, then the amount of memory required is equal to $(2d - 1) \times m$ (for a complete graph, m is equal to $(n(n - 1)/2)$. We will then have, in the worst case, a spatial complexity of around $O(n^2d)$. The worst-case temporal complexity is $O(mnd^3)$.

8.3. Distributed resolution

We present the main methods for the distributed search of solutions to a DisCSP. The majority of authors propose algorithms that process DisCSPs with one variable per agent while drawing from those proposed within the traditional CSP framework. We will, however, discuss, in section 8.3.6, some techniques to go from one variable per agent to several per agent. We assume the following communication model:

– Agents communicate by sending messages. An agent can send messages to other agents if they are its acquaintances.

– The delay in delivering a message is finite. For the transmission between any pair of agents, messages are received in the order in which they were sent.

8.3.1. *Asynchronous backtracking algorithm*

In [YOK 90] and [YOK 92], the authors present the asynchronous backtracking (ABT) algorithm, which adapts the backtracking algorithm with nogood-recording within the distributed framework of DisCSPs. We recall the notion of nogood as inconsistent assignment of values to variables.

Each agent has exactly one variable and tries to instantiate it in a manner consistent with the instantiations of its acquaintances.

To avoid infinite processing loops, such as "changing X_1 leads to changing X_2 which leads to changing X_3 which leads to changing X_1", a total (priority) order relation is introduced among the agents. This order allows the search to be structured. Thus, each agent tries to find a value that satisfies the constraints with the variables of higher-priority agents. When an agent sets a value to its variable, the selected value is sent to lower-priority agents. When no value is possible for a variable, the inconsistency is reported to higher agents in the form of a nogood. ABT is complete, that is it is able to find a solution or detects that no solution exists in a finite time.

The following paragraph, which is essentially drawn from [YOK 95], details the ABT algorithm. Let us make the following assumptions:

– Each agent has exactly one variable.

– All constraints are binary.

– There exists a constraint between any pair of agents.

– Each agent knows all constraint predicates relevant to its variable.

In the following, we use the same identifier X_i to represent an agent and its variable. We assume that each agent (and its variable) has a unique identifier. So, each agent concurrently assigns a value to its variable and sends the value to other agents. After that, agents wait for and respond to incoming messages. There are two kinds of messages: ok? messages to communicate the current value and nogood messages to communicate information about constraint violations. The procedures executed at agent X_i by receiving an ok? message and a nogood message are described in Figure 8.1. An overview of these procedures is given as follows:

– After receiving an ok? message, an agent records the values of other agents in its *agent_view*. The *agent_view* represents the state of the world recognized by this agent.

– The priority order of variables/agents is determined by the alphabetical order of the identifiers, that is preceding variables/agents in the alphabetical order have higher priority. If the current value satisfies the constraints with higher-priority agents in the *agent_view*, we say that the current value is consistent with the *agent_view*. If the current value is not consistent with the *agent_view*, the agent selects a new value that is consistent with the *agent_view*.

– If the agent cannot find a value consistent with the *agent_view*, the agent sends nogood messages to higher-priority agents. A *nogood* message contains a set of variable values that cannot be a part of any final solution.

By using this algorithm, if a solution exists, agents will reach a stable state where all constraints are satisfied. If there is no solution, empty nogood will be found and the algorithm will terminate.

When (OK?, X_j, Nogood) is received do

1. Revise agent-view
2. Check agent-view

When (Nogood, X_j, Nogood) is received do

3. Save Nogood as a new constraint
4. When Nogood contains X_k which is not a neighbor do
5.   Ask X_k to add X_i as a neighbor
6.   Add X_k as a neighbor
7. old-value:= current-value do
8. Check-agent-view
9. When old-value = current-value do
10.   Send (Ok?, (X_j, current-value) to X_j

Procedure Check-agent-view

11. When agent-view and current-value aren't consistent do
12. If it doesn't exist any value into D_i that is consistent with agent-view then
13.   Backtrack
14. Else
15.   Select d $\in$ D_i where agent-view and d are consistent
16.   Current-value:= d
17.   Send (OK ?, (X_i, d)) to neighbors

Procedure Backtrack

18. Generate a Nogood V
19. When V is an empty Nogood do
20.   Inform the other agents that there is no solution
21.   End the algorithm
22. Select (X_j, d_j) where X_j has the littlest priority in the Nogood
23. Send (Nogood, X_i, V) to X_j
24. Delete (X_j, d_j) from; agent-view
25. Check-agent-view

Algorithm 8.1. *Asynchronous backtracking*

8.3.2. *Asynchronous weak-commitment search*

The asynchronous weak-commitment (AWC) algorithm is an improvement of the previous algorithm ABT. It is also inspired by the sequential version of the weak-commitment search for solving CSP.

The main limitation of ABT is the fact that the order of the agents/variables is statically determined and an agent tries to find a value satisfying the constraints with the variables of higher-priority agents. When an agent sets a variable value, the agent commits to the selected value strongly, that is the selected value will not be changed unless an exhaustive search is performed by lower-priority agents. Therefore, in large-scale problems, a single mistake of value selection becomes fatal since doing such an exhaustive search is virtually impossible. On the other hand, in the asynchronous weak-commitment search, when an agent cannot find a value consistent with the higher-priority agents, the priority order is changed so that the agent has the highest priority. As a result, when an agent makes a mistake in value selection, the priority of another agent becomes higher; thus, the agent that made the mistake will not commit to the bad decision, and the selected value will be changed. Thus, agents can revise a bad decision without an exhaustive search by changing the priority order of agents dynamically [YOK 95].

Also note that the AWC algorithm uses the min-conflict heuristic. When selecting a variable value, if there exist multiple values consistent with the agent view (those that satisfy all constraints with variables of higher-priority agents), the agent prefers the value that minimizes the number of constraint violations with variables of lower-priority agents [YOK 95].

As it is known in centralized CSPs, dynamic variable ordering is an efficient heuristic. Recent studies have shown that the same is true for algorithms that perform DisCSPs. As in AWC, these studies suggest heuristics of agent/variable ordering and empirically show large gains in efficiency over the same algorithms using static order. Among them we quote ABT_DO (DO for dynamic ordering) [ZIV 06], DIBT for DIstributed BT algorithm [HAM 98] and interleaved DIBT that improves DIBT [HAM 02b].

8.3.3. *Asynchronous aggregation search*

This method was presented in [SIL 00]. It differs from ABT by the fact that the agents do not exchange messages containing assignments of individual variables but rather messages that include assignments of tuples of variables that remove restrictions in the order of constraint processing. The asynchronous aggregation search (AAS) is based on a technique that allows an aggregation of tuples of values to be propagated instead of individual values, that is there is no sending of single assignments but of Cartesian products of assignments.

Another difference lies in the fact that the agents are no longer responsible for a single variable as before but possibly for several constraints. Furthermore, a new agent, called *system agent*, is added to decide the order of the agents, initialize the links and announce the termination of the search.

Otherwise, the agents are similar to those of ABT, that is they are assigned priorities. A link exists between two agents if they share a variable. The link is directed from the agent with lower priority to the agent with higher priority.

Before presenting the exchanged messages that form the core of the AAS dynamic, we start by giving the necessary concepts end definitions:

An assignment is a triplet (x_j; set$_j$; h_j) where x_j is a variable, set$_j$ a set of values for x_j, and h_j a history of the pair (x_j; set$_j$).

An aggregate is denoted compactly by (V; S; H) where V is the set of variables, and S and H their respective sets of values and histories.

An explicit nogood has the form V, where V is an aggregate.

The view of an agent A_i is as an aggregate (V; S; H) such that V contains variables A_i is interested which in it. Each agent A_i owns a set of local constraints. The variables A_i are interested in those that are implied in its local constraints called the local variables and those establishing links with other agents.

The current solution space of A_i, denoted by C_{Ai}, is described by the local constraints, a list of explicit nogoods and a view.

The agents in AAS communicate via the following messages which are presented in bold:

– **ok?** messages that have an aggregate as a parameter. They represent proposals of domains for a given set of variables and are sent from agents with lower priorities to agents with higher priorities. An agent sends **ok**? messages containing only domains in which the target agent is interested. He does not send domains for assignments for which he was proposed and he has never changed. If he has not discarded a recent applicable nogood, then he sends only the domains for which he proposes a new modification now. **ok**? messages are also sent as answers to **add-link** messages.

– **nogood** messages that have as parameter an explicit nogood. A **nogood** message is sent from an agent with higher priority to an agent with lower priority, namely to the agent with the highest priority among those that have modified an assignment in the parameter. An empty parameter signals failure.

– **add-link** (vars) messages: sent from agent A_j to agent A_i (with $j > i$). They inform A_i that A_j is interested in the variables vars.

Note that AAS provides a natural support for enforcing privacy requirements on constraints. Let us finally note that a quite important part of section 8.3.3 is drawn from [SIL 00].

8.3.4. *Approaches based on canonical distribution*

In [SOL 98], two methods based on canonical distribution of the problem are proposed:

8.3.4.1. *Central first peripheral after*

In this approach, a central solution is first computed by the central agent and then broadcasted to the other agents, called peripheral agents. The latter try (in parallel) to instantiate while respecting the choice of the central agent. If a peripheral agent encounters a dead end, a return is made to the central agent and another central solution is proposed (taking into account the cause of the failure).

8.3.4.2. *Peripheral first central after*

Conversely, in this approach, the peripheral agents first instantiate simultaneously and inform the central agent of their solutions to their sub-problems. The central agent checks the compatibility of these solutions with the central constraints (inter-agent constraints) and sends backtracking messages to the peripheral agents in case of failure.

This distribution suffers from some drawbacks. First, its centralized nature is very pronounced. The central component governs the resolution and may limit the system's degree of parallelism. Furthermore, the canonical approach involves moving certain intra-agent constraints from one agent to another. However, this move is often not desirable (security) or impossible from a practical point of view (data export, specific information on constraints expressed in intention, etc.).

8.3.5. *DOC approach*

8.3.5.1. *The foundations of distributed optimization under constraints (DOC)*

This section is essentially drawn from the PhD thesis of [BEN 11]. Details are also provided in [BEN 08a, BEN 08b]. First of all, let us recall that the external or boundary variables are those that are involved in external constraints (inter-agents) even if they are also involved in internal constraints. The internal variables are those involved only in internal constraints, that is intra-agents. Thus, the boundary of an agent consists of its external variables while its interior consists of its internal variables.

In accordance with the DOC distribution in section 8.1.2, two classes of agent (decision-makers and advisors) are defined. An interface agent was added to act as intermediary between this group of agents and the user to:

– Create and initialize the decision-maker and advisor agents;

– Determine and communicate the order between the agents to ensure, on the one hand, the completeness of the resolution and on the other hand, its speed;

– Launch the resolution process;

– Recognize that the problem relating to the DisCSP was solved by this group of agents;

– Inform the user of the final result.

The order between the agents is determined according to the following five rules:

1) The decision-maker, which has the minimum of advisors, is of the highest priority so as to detect as soon as possible the source of the failure of the global resolution.

2) If two decision-makers (or more) have the same number of advisors, the decision-maker that has the minimum of external variables will have the highest priority.

3) If two decision-makers (or more) remain identical following the application of the first two rules, the highest-priority decision-maker will be the one whose domain size of all external variables is the smallest.

4) Finally, if two decision-makers (or more) have the same order after the first three steps, then the alphanumeric order will be considered.

5) The advisors, whose list of acquaintances contains the highest priority decision-maker, will be of the highest priority.

In what follows, we describe the behavior of the agents within the DOC-SAT framework, the version that deals with the satisfaction of all the constraints for the DisCSP in question.

One of the advantages of DOC is that it does not limit resolution to the use of a very specific algorithm. Indeed; each decision-maker agent can use any CSP resolution and/or consistency reinforcement technique previously described in Chapters 2–4, which allows this approach to be rather flexible and generic.

It is the interface agent that triggers the distributed resolution process by requesting advisors to help their decision-makers to instantiate their internal variables.

8.3.5.2. *The behavior of the decision-maker agents*

Once an advisor has been requested by the interface, it sends an **exploreDirection** message containing a *partial search direction* to each of its decision-makers, in the form of values to be assigned to external variables for which it is responsible (section 8.1.2). When a decision-maker receives all the values concerning its boundary, it launches the process for solving its local problem (local CSP formed by its internal and external variables as well as its internal constraints), using one of the known CSP techniques.

Depending on the result of this resolution process, a decision-maker can have one of two possible states: *satisfied* or *unsatisfied*.

It is satisfied in two cases:

– When its boundary and its interior are consistently instantiated. This means that there is a local solution. To propagate this solution, the decision-maker performs the satisfaction behavior relating to the **exploreDirection** message it received, that is it sends this local solution as *partial direction* by means of the **infoDecision** message to each of its advisors. If it receives an *OK* as a response to the **infoDecsion** message from all of its advisors then it will remain satisfied otherwise it will become unsatisfied.

– When it finishes exploring all possible *partial search directions* sent by all its advisors, its behavior consists in informing the interface agent of its current instantiation and its state of satisfaction.

In the opposite case, i.e. when the decision-maker agent is unsatisfied, it selects one of its advisors not yet requested (sorted according to the order established by the interface agent) and requests that it **change direction**, that is it proposes a new *partial search direction*. Note that when a decision-maker has a single advisor, it requests that it puts the unsuccessful *partial direction*, that is which has led to an unsatisfied state in the set of the *nogood* directions.

Thus, there are three types of message that a decision-maker can receive (see Figure 8.10):

– The first message is a request from one of its advisors, which includes a *partial search direction* to be explored. This message is in the form of **exploreDirection**. This same message can contain the "*Forced*" parameter and in this case, the proposed *partial search direction* must be accepted by the decision-maker.

– The second message is also received from one of its advisors, which includes a *NogoodPartialDirection*. The decision-maker in question then updates its *ExploredSearchDirection* set by removing *search directions* that contain a tuple or a combination of tuples that have become "nogood".

– The third message is a response to the **infoDecision** message.

8.3.5.3. *The behavior of advisor agents*

When the process is triggered by the interface agent, each advisor sends the first element of its *partial search directions* list to each of its decision-makers and awaits the potential arrival of messages. There are two types of message that an advisor can receive (see Figure 8.10):

– The first message is a request for a change in direction called **ChangeDirection** sent by one of its decision-makers.

– The second message is called **infoDecision** that is a request for information regarding the consistency of a partial direction, that is to see if it is consistent or not. This message may contain the parameter *"Forced"* and in this case it becomes an order. In other words, the advisor, which receives this message, tries to find a *partial direction* that contains the tuple received with this message.

As with the decision-maker, an advisor can have one of two possible states: *satisfied* or *unsatisfied*.

– It is satisfied when it no longer receives requests for changes in direction or when all the partial directions received in the **infoDecision** message are consistent. In this case, it sends to all its decision-makers a message to inform them of the consistency of the *partial directions* in question.

– In the opposite case, its unsatisfied behavior consists in choosing a direction that neither belongs to the *taboo* list nor the *nogood* list. The *taboo* list is a list of a fixed-size-containing *directions* that are temporarily forbidden. This list is also dynamically refreshed to diversify the search as much as possible. If there is no longer a direction that satisfies this condition (neither belonging to the *taboo* list nor the *nogood* list) then it will randomly select a *nogood partial direction*, which is not in the *taboo* list. When an advisor receives an **infoDecision** message, its unsatisfied behavior consists in choosing the decision-maker that maximizes the conflict with its *partial direction* and sending it a message containing a response and a new *partial direction* that is also communicated to the other decision-makers.

Depending on the nature and number of constraints that an advisor supervises, a *partial direction* can correspond to a tuple of values or a combination of tuples.

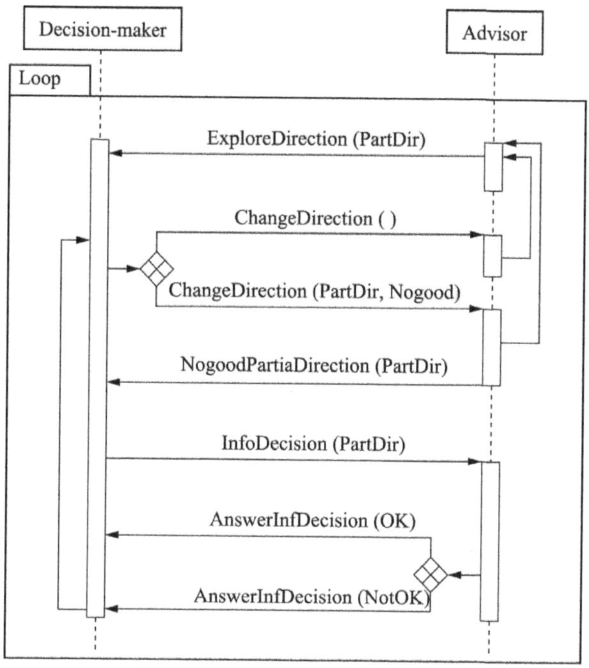

Figure 8.10. *Sequence diagram (DOC communication protocol)*

8.3.5.4. *Extensions of DOC*

Several studies have extended the basic version of DOC-SAT, for DisCSPs where it is a matter of satisfying all the internal and external constraints to DCSOP. These studies gave rise to DOC-OptC for **C**entralized **DOC-Opt**imization, which consists of a single objective function that involves all the variables of all decision-maker agents. It is the interface agent that is responsible for this [BEN 09a, BEN 09b].

The second extension is DOC-OptD for **D**istributed **DOC-Opt**imization. Indeed, as its name indicates, optimization is fully distributed at the level of the decision-maker agents. Each decision-maker thus has its own objective function that it autonomously and locally optimizes while, of course, cooperating with other agents. It should be noted that these agents do not necessarily use the same optimization method [BEN 09b].

Heuristics have also been proposed in this framework so as to accelerate the search while improving the objective functions [BEN 08c].

8.3.6. *Generalization of DisCSP algorithms to several variables*

In [YOK 98], the authors show that algorithms solving DisCSP with one variable per agent can be applied to the case of several variables per agent, via the use of one of the two following methods:

8.3.6.1. *Method 1*

Each agent seeks all the solutions to its local problem. The tuple of all the variables relative to its local problem will be considered as a meta-variable having as a domain the set of all these solutions. Thus, the whole problem can be formulated as a DisCSP where each agent will be responsible for its meta-variable. The drawback of this method is that when the local problem becomes complex, finding all the solutions becomes impractical.

8.3.6.2. *Method 2*

An agent creates multiple virtual agents; each of them is responsible for a single local variable and then simulates the activity of these virtual agents. In this case, each agent (real) does not have to determine all local solutions as before. The drawback of this method is the expensive communication between the virtual agents and the real ones.

Note that DOC and AAS can handle agents with several local variables. In [YOK 98], variable-ordering AWS, an extension of the AWC algorithm that enables several local variables per agent, was developed. As its name indicates, it introduces a dynamic total order (priority) relation on local variables.

In [ARM 97], the assignment of priorities between the agents to handle several local variables was introduced. In the underlying algorithm, called agent-ordering AWS, each agent tries to find a local solution that is consistent with the local solutions of the higher-priority agents. If there is no such solution, there is either backtracking or a change in priorities. This approach is, in fact, similar to the first method described earlier, with the exception that each agent only seeks local solutions if necessary, instead of finding them in advance.

8.4. Bibliography

[ARM 97] ARMONSTRONG A., DURFEE E., "Dynamic prioritization of complex agents in distributed constraint satisfaction problems", *Proceedings of the 15th International Joint Conference on Artificial Intelligence (IJCAI-97)*, San Francisco, Morgan Kaufmann Publishers, pp. 620–625, 1997.

[BEN 02a] BEN HASSINE A., GHÉDIRA K., "How to establish arc-consistency by reactive agents", *Proceedings of the European Conference on Artificial Intelligence*, Lyon, France, 2002.

[BEN 02b] BEN SALAH K., GHÉDIRA K., "Foundations of DOC-sat", *Proceedings of the Workshop on Distributed Constraint Reasoning* in conjunction with *Autonomous Agents and MultiAgent Systems*, Bologna, Italy, 2002.

[BEN 08a] BEN SALAH K., GHÉDIRA K., "An efficient method for distributed constraint satisfaction problems resolution", *The European Multidisciplinary Society for Modelling and Simulation Technology-Industrial Simulation Conference, EUROSIS-ISC*, Lyon, France, June 2008.

[BEN 08b] BEN SALAH K., GHÉDIRA K., "DOC: a new multi-agent approach for DCSP resolution", *NOTERE 2008, 8th Annual International Conference on New Technologies of Distributed Systems*, Lyon, France, June 2008.

[BEN 08c] BEN SALAH K., GHÉDIRA K., "Using heuristics to optimize the research in a distributed environment", *EUMAS 2008*, Bath, UK, December 2008.

[BEN 09a] BEN SALAH K., GHÉDIRA K., "DOCopt a new method for distributed constraint satisfaction and optimization problems resolution", *IADIS International Conference Intelligent Systems and Agents 2009*, Algarve, Portugal, 2009.

[BEN 09b] BEN SALAH K., GHÉDIRA K., "Using MAS to solve distributed constraint satisfaction and optimization problems", *ICAI'09 - The 2009 International Conference on Artificial Intelligence*, Las Vegas, NV, 2009.

[BEN 11] BEN SALAH K., DOC: une approche générique Distribuée d'Optimisation et de satisfaction de Contraintes, Doctoral Thesis, ENSI, Tunisia, 2011.

[BOW 06] BOWRING E., TAMBE M., YOKOO M., "Multiply constrained distributed constraint optimization", *AAMAS 2006, Autonomous Agents and Multi-Agent Systems*, Hakodate, Japan, 8–12 May 2006.

[BRI 09] BRITO I., MEISELS A., MESEGUER P., ZIVAN R., "Distributed constraint satisfaction with partially known constraints", *Constraints*, vol. 14, no. 2, pp. 199–234, June 2009.

[COL 99] COLLIN Z., DECHTER R., KATZ S., "Self-stabilizing distributed constraint satisfaction", *Chicago J. Theor. Comput. Sci.,* 1999.

[COO 88] COOPER P.R., SWAIN M.J., "Domain dependence in parallel constraint satisfaction", *Proceedings of the National Conference on Artificial Intelligence*, Los Altos, CA, USA, IOS Press, 1988.

[FAL 08] FALTINGS B., LÉAUTÉ T., PETCU A., "Privacy guarantees through distributed constraint satisfaction", *International Conference on Web Intelligence and Intelligent Agent Technology 2008*, IEEE/WIC/ACM, Sydney, NSW, Australia, 9–12 December 2008.

[FKI 02] FKI W., GHÉDIRA K., "Forward-checking self-stabilizing protocol for distributed constraint satisfaction", *The 3rd International Workshop on Distributed Constraint Reasoning at AAMAS'02*, Bologna, Italy, 2002.

[GRE 07] GREENSTADT R., Improving privacy in distributed constraint optimization, PhD Thesis in computer science, Harvard University, Cambridge, MA, June 2007.

[HAM 98] HAMADI Y., BESSIERE C., QUINQUETON J., “Backtracking in distributed constraint networks”, *Proceedings of the European Conference on Artificial Intelligence,* Brighton, UK, pp. 219–223, 1998.

[HAM 02a] HAMADI Y., “Optimal distributed arc-consistency”, *Constraints*, vol. 7, no. 3–4, pp. 367–385, 2002.

[HAM 02b] HAMADI Y., “Interleaved backtracking in distributed constraint networks”, *International Journal on Artificial Intelligence Tools*, vol. 11, no. 2, pp. 167–188, 2002.

[HAN 00] HANNEBAUER M., “On proving properties of concurrent algorithms for distributed CSP”, *Proceedings of the Workshop on Distributed Constraint Satisfaction* in conjunction with *the International Conference on Principles and Practice of Constraint Programming*, Singapore, 2000.

[HIR 97] HIRAYAMA K., YOKOO M., “Distributed partial constraint satisfaction problem”, in SMOLKA G. (ed.), *Proceedings of the International Conference on Principles and Practice of Constraint Programming*, pp. 222–236, Schloss Hagenberg, Austria, 1997.

[HIR 00] HIRAYAMA K., YOKOO M., “An approach to over-constraint distributed constraint satisfaction problems: distributed hierarchical constraint satisfaction”, *Proceedings of the International Conference on Multi-Agent Systems*, Boston, MA, USA, 2000.

[KAT 06] KATIA S., CHECHETKA A., “No commitment branch and bound search for distributed constraint optimization”, *AAMAS*, Hakodate, Japan, 8–12 May 2006.

[MAC 85] MACKWORTH A.K., FREUDER E.C., “The complexity of some polynomial network consistency algorithms for constraint satisfaction problem”, *Artificial Intelligence*, vol. 25, no. 1, pp. 65–74, 1985.

[MOD 02] MODI P.J., SHEN W.M., TAMBE M., YOKOO M., "An asynchronous complete method for general distributed constraint optimization", *The 3rd International Workshop on Distributed Constraint Reasoning at AAMAS'02*, Bologna, Italy, 2002.

[NGU 95] NGUYEN T., DEVILLE Y., "A distributed arc-consistency algorithm", *Proceedings of the International Workshop on Concurrent Constraint Satisfaction*, Venise, Italy, May 1995.

[PEA 07] PEARCE J.P., TAMBE M., "Quality guarantees on k-optimal solutions for distributed constraint optimization problems", *IJCAI, Twentieth International Joint Conference on Artificial Intelligence*, Hydrabad, India, 6–12 January 2007.

[SAM 87] SAMUEL A., HENDERSON T., "Parallel consistent labeling algorithms", *International Journal of Parallel Programming*, vol. 16, pp. 341–364, 1987.

[SIL 00] SILAGHI M.C., SAM-HAROUDS D., FALTINGS B., "Asynchronous search with aggregations", *Proceedings of* the *National Conference on Artificial Intelligence*, Austin, Texas, USA, AAAI Press, pp. 917–922, 2000.

[SIL 08a] SILAGHI M.C., DOSHI P., MATSUI T., YOKOO M., "Distributed private constraint optimization problem: cost of privacy loss", *Distributed Constraint Reasoning Workshop (DCR)*, Estoril, Portugal, May 2008.

[SIL 08b] SILAGHI M.C., YOKOO M., "ADOPT-ing: unifying asynchronous distributed optimization with asynchronous backtracking", (electronic version has appeared in November 2008) *Journal of Autonomous Agents and Multi-Agent Systems (JAAMAS)*, vol. 19, no. 2, pp. 89–123, November 2008.

[SOL 98] SOLOTOREVSKY G., GUDES E., MEISELS A., Distributed constraint satisfaction problems – a model and application, BGU, Technical Report, 1998.

[SWA 88] SWAIN M.J., COOPER P.R., "Parallel hardware for constraint satisfaction", *Proceedings of the National Conference on Artificial Intelligence*, Los Altos, CA, USA, IOS Press, pp. 682–686, 1988.

[YOK 90] YOKOO M., IDHIDA T., KUWABARA K., "Distributed constraint satisfaction for DAI problems", *Proceedings of the The International Workshop on Distributed Artificial Intelligence*, Texas, USA, October 1990.

[YOK 92] YOKOO M., DURFEE E.H., ISHIDA T., KUWABARA K., "Distributed constraint satisfaction for formalizing distributed problem solving", *Proceedings of the International Conference on Distributed Computing Systems*, Yokohama, Japan, IEEE Computer Society Press, pp. 614–612, June 1992.

[YOK 93] YOKOO M., "Constraint relaxation in distributed constraint satisfaction problem", *Proceedings of the International Conference on Tools with Artificial Intelligence*, Boston, Massachusetts, pp. 56–63, 1993.

[YOK 95] YOKOO M., "Asynchronous weak commitment search for solving distributed constraint satisfaction problems", *Proceedings of the International Conference on Principles and Practice of Constraint Programming*, Cassis, France, LNCS, Springer, 1995.

[YOK 96] YOKOO M., HIRAYAMA K., "Distributed breakout algorithm for solving distributed constraint satisfaction problems", *Proceedings of the International Conference on Multi-Agent Systems*, Kyoto, Japan, AAAI Press, pp. 401–408, 1996.

[YOK 98] YOKOO M., HIRAYAMA K., "Distributed constraint satisfaction algorithm for complex local problem", *Proceedings of the International Conference on Multi-Agent Systems*, Paris, France, 1998.

[ZIV 06] ZIVAN R., MEISELS A., "Dynamic ordering for asynchronous backtracking on DisCSPs", *Constraints*, vol. 11, pp. 179–197, 2006.

Index

D

E

F

G

H, I

K, L

M

N, O

P

S

T, U

V, W

Printed and bound by CPI Group (UK) Ltd, Croydon, CR0 4YY
13/09/2026
14961081-0002